The most beautiful

L - Numbers

Ulrich Glaser sen.

Acknowledgements

Photos:

We would like to thank the following specialists, companies, breeders and hobbyists for their advice and kindly letting us use their slides. We also thank all those we might have forgotten.

Ernst Otto von Drachenfels **Hans-Georg Evers**
Kerstin Holota **Karl Lang †**
Uwe Römer **Erwin Schraml**
Christoph Seidel **Ingo Seidel**
Frank Teigler **Frank Warzel**
Uwe Werner **Archiv A.C.S. Nakano**

Aquarium Glaser GmbH:
for providing beautiful fish for our photographers from their weekly imports

amtra - Aquaristik GmbH:
for providing furnished aquaria and equipment for testing

Veterinary consultant:
Dr. med. vet. Markus Biffar,
veterinarian, fish specialist

Liability:

All the information in this book has been recorded with due care and attention. The authors and the publishers will not accept liability for any inaccuracies.
By purchasing this book the owner explicitly accepts this disclaimer of liability.

All rights reserved. Except for private study, research or review no part of this publication may be reproduced, stored in any retrieval system or reproduced in any form by any means without the permission of the publishers.

Further useful tips about care and maintenance can be found every six weeks in AQUALOGnews, the unique newspaper for all friends of the hobby.

Read, for example, the latest breeding reports in the news. It is available in German or English and can be obtained at your local pet shop or subscribed to at the publisher.
Order your free specimen copy!

Further literature references on page 47 in the back of this book.

AQUALOG: *Special* - Serie Ratgeber
Rodgau: A.C.S.
The most beautiful L-numbers - 1998

The most beautiful L-numbers
Ulrich Glaser sen. - Mörfelden-Walldorf: A.C.S. (Aqualog)

ISBN 3 - 931702 - 33 -2
NE: Glaser, Ulrich sen.

© **Copyright by:**   Verlag A.C.S. GmbH
Liebigstraße 1
D-63110 Rodgau
Germany

Author:
Ulrich Glaser sen.
Scientific consultant:
Frank Schäfer, Dipl. Biol.
Translation:
Monika Schäfer, M.A.
Index und organisation:
Wolfgang Glaser
Editor:
Frank Schäfer, Dipl. Biol.
Cover layout:
Gabriele Geiß, Frankfurt/Main

Print, typesetting, processing:
Lithographics: Verlag A.C.S.
Prepress/Photo processing: Frank Teigler, Michael Blügell, Dipl. Wirt. Ing. (FH)
Print: Giese-Druck, Offenbach
Printed on EURO ART,
100 % chlorine free paper

Editors address:

Verlag A.C.S. GmbH
Liebigstraße 1
D-63110 Rodgau
Fax: +49 (0) 61 06 - 64 46 92
e-mail: acs@aqualog.de
http://www.aqualog.de

PRINTED IN GERMANY

Cover Photos

Front cover:
L 91 *Leporacanthicus triactis*
- E. Schraml/Archiv A.C.S.

Contents

	page:		page:
The Author	6	**Maintenance:**	
Preface	7	The right foods	28
		Feeding	29
Basics:		Food and roots	30
What are catfishes?	8	Disease prevention	31
What are armoured catfishes?	9	The settling-in periord	32
What are L - numbers?	10	The partial water change	33
Why L - and LDA - numbers ?	11	Disease treatments	34
L - numbers and their trading names	12		
Distribution and special characterstics	13	**Breeding:**	
Habitats and anatomical features	14	Basic requiremets	37
		Tips and breeding reports	38
The suitable tank for loricariids:		Rearing the young	39
The right tank	15	Differences in the sexes	40
The right water	16		
Setting up the new aquarium	18	**The AQUALOG system**	
The right plants	19	Information and description	42
		Outlook: Future L-numbers ?	44
Community tank with different species:		Photos: New L-numbers on the waiting list	45
Basic rules	22		
Behaviour towards other fishes	23	Index	46
		Index and Literature references	47
Choosing fishes for your tank - suggestions	26	Key to the symbols	48

The author

Ulrich Glaser sen.

He spend his youth (Mr. Glaser was born in 1937) trying to get through the awful times following WW II as best as he could. Of course, his mother was concerned with getting herself and her family through these times of poverty and hunger and there was neither time nor opportunity to get interested in any kind of 'hobby'. Taking up studies at university was simply impossible.

Owing to the circumstances, Ulrich Glaser's first contact with the aquarium hobby was finally made in a pet shop. There, he learned absolutely anything one needs to know about keeping ornamental fishes, from cleaning tanks to breeding rare and precious species. Today, he is looking back on years and years of aquarium and fish tending - a very experienced enthusiast he is indeed.

After having managed several ornamental fish wholesale companies, he founded his own business, together with his wife and his two oldest children, in 1984: Aquarium Glaser, today Europe's biggest ornamental fish import/export company. The firm is now managed by his daughter and has gained a lot of respect in the ornamental fish business.

A few years later, a second branch developed from the fish wholesale company: amtra-Aquaristik, managed by his older son. amtra-Aquaristik produces the well-known amtra products which are sold world-wide.

Still being involved in the import/export business, Ulrich Glaser often found it extremely difficult, if not impossible to identify newly imported fishes - there simply was no useful identification literature available. Out of this unsatisfactory situation the idea developed to start an identification catalogue series: the AQUALOG was born.

In 1995, Ulrich Glaser founded Verlag A.C.S., together with his younger son. The biologist Frank Schäfer, taking over the task of scientific editor, completed the AQUALOG team.

Our team has set itself the goal to catalogue all known ornamental fishes of the world and publish reference books with high-quality, multi-coloured photographs. There are - approximately - 40000 fish species and we have a long way to go if our task is to be realised. But we think, friends of the hobby and specialists alike will appreciate our work.

In order to keep the AQUALOG series handy and easily comprehensible, each book treats one group or genus of fishes. The most beautiful and popular species are shown on an extra poster and are introduced in this Special series. All species belonging to each group or genus, including all varieties and breeding forms, are presented in the respective AQUALOG book.

All AQUALOGs can be supplemented which means that pictures of newly discovered or bred fishes can be stuck in the free pages in the back part of the book. This way, the owner of an AQUALOG can - easily and cheaply - keep the catalogue up-to-date for years and years.

Also, we developed a genuine code-number-system that labels every single fish species and its varieties with an individual code-number. The fish keeps this number even if some day its name is changed. The individual code-number makes every fish absolutely distinctive and internationally communicable, leaving language barriers behind.

Wolfgang Glaser

Preface

Travel-mad aquarists are always on the look-out for new information about the natural environment and living conditions of their pets. When examining the water, they often discover species that were unknown to either the aquarium hobby or even science. The local fishermen closely observe everything the tourists do so that they get an idea which species are popular in Europe. On the photo you can see a beautiful L-021.

The AQUALOG advisory special series provides easily comprehensible information, presenting fishes of a certain group in many beautiful pictures, easy text and useful drawings.

All aquarists and especially beginners who start to get involved in the wonderful hobby, will find a lot of useful advice regarding care, maintenance and breeding.

The advisory specials are written by experienced specialists who love to tell you all about the tips and tricks one can learn when spending many years keeping and breeding ornamental fishes.

This second issue of the specials series is dedicated to the armoured catfishes, especially to the so-called L-numbers.

Why these fishes are called L-numbers, where they come from and how the strange name came about, the proper care and with which other species they can be kept together is explained in great detail.

Breeding armoured catfishes is not that easy, one has to spend a lot of time watching their behaviour. Yet, some species of the group called L-numbers have been successfully bred in the past, a sign that they feel well in the aquarium.

Reports about such successful breeding are included in this advisor: One of the breeders tells you all the tricks for the repetition of this success.

Besides the pictures included in the text, we present the 64 most beautiful and popular species on the accompanying poster. The captions not only provide the L- or LDA-number of the fish shown in the photo but also its common trading name and, if the fish has been scientifically described, its Latin name.

Further, the captions include care instructions and the most important characteristics of the respective fish in form of catchy symbols so that you don't have to re-read the text in order to get the basic information about a certain fish. The explanation of the symbols is on page 2 of this book.

We are especially grateful to all photographers and all others who were involved in the making of this book and who provided their knowledge.

We hope you enjoy reading this book and that it will help you with the successful keeping of these wonderful fishes.

Your AQUALOG team

Wolfgang Glaser

Basics:
What are catfishes ?

The catfishes, of the order Siluriformes, are with an estimated number of 3 500 species one of the largest groups of fishes among the about 40 000 (this is an estimated number, including undescribed species) fish species that exist world wide. The order of catfishes is divided into families. One of these families is the group of loricariids, or suckermouth or armoured catfishes.

It is very hard to give a general overview of this vast group of fishes, owing to the extreme large number of species belonging to it and the fact that they are distributed over the whole planet. Still, it can be said that the majority of these fishes lives in the tropical and subtropical regions of the earth.

Because catfishes possess a lot of interesting features, they are precious objects of scientific studies and offer a lot of valuable information. The numerous different - and sometimes bizarre - shapes and looks of these fishes make them, and especially the dwarf species, popular pets in the aquarium hobby. In the past years, many enthusiasts have turned their interest completely towards the care and breeding of catfishes.

When thinking of a catfish, many people imagine a slimy fish without scales, a big head and feeler-like barbels. But nature has brought forth a whole lot of other forms, many of which inhabit the most different parts of the earth, from deep river valleys to high mountain regions.

In the course of time, the catfishes have adapted to their environment, developing in the process the different forms we know today. Some have a bony plated body, some use their sucker mouths to attach to the substrate in the rapidly flowing waters they live in; others search the muddy grounds of their swamp-like habitats for food, using skilfully their extremely long barbels.

There are many more curious forms with interesting features like the electric catfishes which are able to generate an electrical shock (you cannot touch them with your bare hands without being hurt), or the glass cats that are more or less transparent, or the upside-down cats with their funny way of swimming. But these catfishes will be written about in another book, because they do not belong to the armoured cats we talk about in this volume.

Most catfishes are bottom dwellers. The colour of most catfishes is a sign for the fish's perfect adaptation to their environment - and also a reason for being considered boring because not colourful. So it is for their interesting behaviour that many hobbyists still are fascinated by these fishes.

The armoured catfishes with over 600 scientifically described species are the largest family within the order of catfishes. I think, at the moment, nobody can claim to know how to exactly systematise the armoured cats.
Science is still working on the compilation of a sensible systematology of this huge group of fishes.

The so-called "L-numbers" also belong to this group; the scientific name of this family is "Loricariidae".

The family of Loricariidae is divided into several sub-families. They are: Ancistrinae (with, for example, the species *Ancistrus, Peckoltia, Baryancistrus, Hypancistrus)*, Hypostominae (e.g. *Cochliodon, Hypostomus, Glyptoperichthys, Liposarcus, Pseudorinelepis, Aphanotorulus*) and Loricariinae (with, e.g., *Loricaria, Ctenoloricaria, Rineloricaria, Loricariichthys*).

What are L-numbers?

So, why is there the separate group of the so-called L-numbers within the family of armoured catfishes?

Some species were imported to Germany as early as the turn of the century, in the very beginning of the hobby, like, for example, the Dwarf Sucker Cats of the genus *Otocinclus*, plecos of the genus *Hypostomus* and *Glyptoperichthys*. The Bristlenoses of the genus *Ancistrus* have been successfully bred for quite a while now, too.

Basics:
What are armoured catfishes

As early as in 1894, armoured cats of the genera Hypostomus *(they were called* Plecostomus *at that time) and* Loricaria *were imported for the aquarium hobby, like the drawing by K. Neunzig proves. Then, L-numbers were, of course, no subject of discussion. Top right:* Corydoras nattereri

Hobbyists very much like to have them in their aquaria, not only for their interesting behaviour, but also because these cats eat algae and food left-overs: they keep the tank clean and are, thus, "useful".

The reason for the late discovery of the colourful catfish species available today under name "L-numbers" might be found in the mentality of the native collectors. These men collect ornamental fishes as often as possible from the same patch of water, near the place where they live and (most conveniently) at the same place where they catch their daily meals.

This is, because, in the first place, the Indios regard all fish as food. And a large Royal-Blue-Discus does not only look nice in a frying pan, it also tastes delicious! Small tetras (like the Cardinal Neon) can still be used for soup, a specimen of L-025 or L-047 does nicely on the barbecue. This is natural and sensible - in Europe, nobody would think about having a trout or cod in the aquarium and be distressed about eating it.

Accordingly, the Indios only smile pityingly when they are asked to be really careful when catching the cats so that the 'pets' are delivered intact and healthy. Also, one has to consider that collecting the colourful armoured catfish species is an exhausting and uneconomical business. While the 'early' species were easy to catch because they live in easily accessible biotopes with sandy bottoms (*Rineloricaria*), in patches of water plants (*Otocinclus*), near river banks (Plecos) or in small streams (*Ancistrus*), most of the colourful species live isolated in between layers of large stones or driftwood.

In these biotopes, catching is tiresome, each single animal has to be collected by hand, sometimes even diving equipment has to be used. No sensible human being takes this trouble without having to.

But now, the world is getting 'smaller' and long-distance flights are getting cheaper and faster by the day.

Enthusiasts are able to travel to foreign countries and visit the tropical places "their" fishes come from.

The native collectors follow the foreign "doctors'" trails and learn which species are popular and much sought-after; the comparatively high prices that are paid for the rarer species are certainly the main reason for some Indios to invest more time and work and hunt for the colourful species that burst on the European market in the late 80s.

Basics:
Why L- and LDA-numbers?

With this catfish, everything started: In DATZ 12/88 the first fish was labelled with an L-number, L-001. In 1991, this catfish from Tocatins/Brazil was classified by Weber as Glyptoperichthys joselimaianus.

Until then, no-one would thought such an amazing array of colours possible in catfishes: with for example, an all-over yellow spot pattern or orange fins or zebra-like stripes like L-046, *Hypancistrus zebra*. The 'new' species were a sensation and triggered off a real "cat-boom".

The rising demand for colourful catfish species stimulated collectors to search for more and more of these exciting fishes - and they found them! Of course, as many colourless species as colourful ones were found, but they never really established themselves in the trade. During this first big wave of imports, it never was quite clear whether these species were only 'new' to the hobby or also 'new' to science. To be able to get, at least, some order into the chaos, Rainer STAWIKOWSKI had the brilliant idea to simply number all new imports until there'd be time to classify them properly.

To make clear that the numbers marked new Loricariidae species, he put an "L" in front of the number - the L-numbers were born. Being the chief editor of the German publication DATZ, he had the perfect media at his hands to publish all new species immediately and present them to enthusiasts as soon as they were imported.

At the time, Rainer STAWIKOWSKI probably didn't suspect the 'cat-mania' that was about to follow. Every single freshly caught or newly imported loricariid was now closely examined and each spot or stripe that was different from a known pattern or the fact that the fish had been collected at a previously unknown habitat was reason enough to label the fish with a new L-number.

It is important to remember that in the late 80s, only very few species were known and that only very small numbers of these fishes were marketed at incredibly high prices.

Basics:
Why L- and LDA-numbers?

The amazing variety of coloration within one species that is displayed especially by suckermouth cats was not yet discovered and recorded. And last but not least, some 'specialists' thought this to be the perfect situation to make a mark on the scene. As a result, photos of the same species, with different colorations like night- or shock-coloration, juvenile and adult coloration were presented - all of them with different L-numbers. One has to admit that in loricariids young and adult fish really differ considerably from each other in appearance. But also already known species, like the fishes described in Isbrücker's 1980 catalogue, were labelled with an L-number - thorough enquiries (which are indispensable) were not made. The often very hectic situation in an editorial office was another reason for some (unnecessary) mistakes, like mixing up photos or printing incorrect captions. Errors like these were (most of the time) discovered sooner or later, and corrected. But still - today, more than 320 L-numbers are known and this is way too much to be comprehensible.

Unfortunately, there is no end in sight. Adding to the already confusing situation was the fact that in Brazil (in the spirit of competition?) the same species were labelled with so-called TR-numbers which has been (fortunately) discontinued.

Similarly, another German periodical, "Das Aquarium", presents catfishes with LDA-numbers. In the past, very often the same fish got an L- as well as an LDA-number. This confusing situation came about not (so we were told) because the DATZ labels loricariids with L-numbers, but Germany's biggest importer who cannot possibly know which new species might already have been imported by other wholesalers. For hobbyists, though, this chaotic situation is not as bad as one might think it would be. To enthusiasts who keep these beautiful fishes in their tanks at home, these are minor matters. In order to give hobbyists the opportunity to get an overview of the current situation, AQUAOLG "Loricariidae: All L-numbers" catalogued all known L- and LDA-numbers and labelled them with a code-number, so that names could be left out of consideration if necessary. Since its first publication, many supplementary stickers with newly discovered or described L- and LDA-numbers have been published; thus, the book is still up-to-date.

Right in the middle of the jungle, collecting ponds were build. Such improvised installations have to be set up when ornamental fishes are collected far away from 'civilisation'. Of course, this kind of 'luxury' is only possible if the dealers are willing to pay for the extra work and effort.

Basics:
L - numbers and their trading names

Due to the completeness of the catalogue, the incredible number of varieties in one species could be documented, also all 'lookalikes' that encourage false identification. With this book, hobbyists can be quite sure to identify their suckermouth cats correctly - as correct as one can identify a species with the help of a photograph.

Beside the number and scientific name, most species have also a common or trade name like

L-002 = Tiger-Pleco,

L-014 = Goldy-Pleco,

L-025 = Scarlet-Pleco,

L-030 = Peppermint-Pleco,

L-047 = Magnum-Orangeseam-Pleco,

L-198 = Zombie-Pleco etc.

These trade names origin from the importers' need to label the fishes somehow before a scientific name or an L-number is available. The names are often quite descriptive but they are of no scientific value.

Classifications like "*Pleco*", "*Peckoltia*" or "*Ancistrus*" are not made on a scientific basis but purely from an onlooker's point-of-view, according to typical features in appearance.

Most of the time, the name used by the exporter on the stock list is simply adopted, like in the cases mentioned above.

But sometimes, these names have really funny origins like in the following example:

L-137, now known as *Cochliodon* cf. *cochliodon* (cf. means "confer", which is Latin and can be translated as "compare with". It is used when a determination is not absolutely safe) is also known as "Violet Red Bruno".

The fish got this name long before the L-numbers were introduced.
When the first specimens of this purplish/reddish-coloured species arrived from Paraguay at the tanks of a well-known Dutch importer it was completely unknown to the crew that unpacked the freight.

Even the consultation of specialist literature didn't have any result. But the fish had to be named one way or the other and when the members of the crew looked at one of their colleagues, Bruno with the copper-red hair, they called the fish (funny as Dutch people are) "Violet Red Bruno". And this name stayed with the fish ever since...

The high numbers of new species and the import quantities soon alarmed environment-conscious people who saw the danger of exploiting the localities.

But luckily, there are, at the moment, no signs of overfishing. Anybody who has seen these habitats with his own eyes will have to admit that overfishing is simply not possible.

Still, the natural environment of loricariids (and other species) is endangered by all kinds of pollution and destruction, like the use of highly poisonous pesticides, fire clearing and gold mining.
Today, the only argument that can convince local governments to pursue an ecological policy is the fact that exporting ornamental fishes is a secure and profitable business.

But so is gold, and gold-panning is a pest that spreads all over South America and pollutes rivers with mercury - whole ecosystems are destroyed or close to an environmental catastrophe.
Reports by aquarists who travelled several times to the same locality are devastating.

Basics:
Distribution and special characteristics of L- numbers

All armoured catfishes come from South America. A vast majority of the known species are endemic to Brazil, but there others that live in Venezuela, Peru, Ecuador, Paraguay, and the tropical parts of Bolivia. Only very few species are found in subtropical regions.

Like already mentioned, all L-numbers belong to the family of armoured cats and, with it, to the following subfamilies:

Ancistrinae with the genera*:

Ancistrus: L-034, L-059, L-071, LDA-002, LDA-005
Chaetostoma: L-188, LDA-011
Baryancistrus: L-018, L-057, L-081
Lasiancistrus: L-168
Hopliancistrus: LDA-015
Hypancistrus: L-046, L-098
Panaque: L-002, L-090, L-190, L-204, LD-0 22
Parancistrus: L-030
Peckoltia: L-005, L-006, L-070, LDA-019
Scobinancistrus: L-014, L-048
Leporacanthicus: L-007, L-091
Lithoxus: L-052
Megalancistrus: L-234
Oligancistrus: LDA-014
Pseudacanthicus: L-024, L-065, LDA-007

Hypostominae with the genera*:

Cochliodon: L-050, L-137, L-238, LDA-021
Glyptoperichthys: L-022
Hypostomus: L-011, L-078, L-166, L-241, LDA-009
Pterygoplichthys: L-154
Pseudorinelepis: L-095

Loricariidae with the genera*:
Rineloricaria: L-010a

(* In this list only genera with L-numbers are included. Of course, there are countless others.)

All armoured cats that have been labelled with L-numbers so far, are bottom-dwellers. This means that they live almost exclusively on the ground of their home waters, where they search for food.

Some species have 'antennas' (barbels), some even a real 'moustache' or 'whiskers' (like cats - that's where these fishes got their name from !).

Others possess a more or less large sucker mouth with which they rasp off algae or (in carnivorous species) attach themselves to stones, roots or driftwood in rivers with a strong current.

Others, like, for example, L-190 *Panaque* sp., have spoon-like teeth in their sucker mouths. These enable them to steadily rasp off the surface of the pieces of wood they cling to; driftwood is simply everywhere in their jungle biotope and one of the cats' main food sources.

All the adaptations described above, of course, are the result of a very long development. In the course of time, the suckermouth cats have adapted to their habitats - whether they live in shallow, rapidly flowing streams, big rivers, swamp-like side arms, stagnant waters or flooded areas where they might even have to survive in small pools during the dry season.

Thus, the way these cats look very often allows to draw conclusions about the biotope the respective species lives in. This again is very important for the setup of a loricariid tank, a topic we will come back to later in the book.

Now, the barbels are a very important physical feature. They possess sensory receptors and are thus an organ for feeling and tasting. One can compare them to the human tongue which is also an organ for tasting and feeling.

As the cats cannot see very well, the barbels are even more important, although the substrate-oriented way the cats live doesn't require good eyes.

Most armoured catfish species can spread out their fins when threatened. In many species, the fins have sharp spines; Ancistrinae even possess so-called interopercle odontodes, sharp spines near the gills that can be everted. When threatened or attacked, the fish can spread these sharp spines, thus having a perfect defensive weapon - with the everted spines it is almost impossible for predators to swallow the small cats.

Basics:
Habitats and anatomical features

When threatened, Ancistrus species spread their interopercle odontodes for defence (like the L-110, right). Usually, these sharp spines are hidden partly in special folds in the skin. This can be clearly seen in the picture of the Ancistrus sp. from Roraima, Brazil (left). The fish is no L-number.

This photo of a loricariid biotope was made in British Guyana.

Armoured catfishes live very often in such rocky brooks. The weapon on the back of our brave collector hints towards the sometimes really dangerous conditions under which some species are collected.

View of the Rio Paraguay. From this river system, many beautiful ornamental fish species are collected, including L-numbers. Small photo: Collectors at work in the Rio Paraguay. To think that such collecting methods could reduce or even endanger a certain fish species, is simply absurd

The suitable tank for Loricariids
The right tank

All so-called L-numbers are, like already mentioned, bottom-dwellers or, more precisely, substrate-oriented fishes.

They need to have body contact with the ground or some other kind of surface.

There are species that stay relatively small, like cats from *Ancistrus* or *Peckoltia* group, and others that grow up to 1 m long, like, for example, Pseudacanthicus.

Of course, one cannot claim the smaller species to be exclusively suitable for aquatic purposes. Hobbyists always have to stick to one simple rule: the larger the tank, the easier the fish-keeping. Today, many enthusiasts are prepared to furnish their homes with tanks as large as 5 or 6 ms length. Such huge tanks are indeed fascinating, if you have the space to install them.

Just imagine, one of your living room walls being completely made of glass, behind it a well-lit, beautifully arranged underwater landscape. Watching such a scenario takes you away into another world - it might even compensate for a holiday.

Well, whether small or large tank - it can be delightful anyway and one has to decide which is the right one, appropriate for the individual taste, room to spare and money to spend.

For information about the required tank size and other basic care instructions, please read the captions accompanying the photographs.

Usually, armoured catfishes are tended in a so-called "South America tank". Smaller species can be kept in community with tetras which inhabit the upper and middle regions of the tank.

Larger species should be kept in aquaria with river-like characteristics, in community with large tetras and/or quiet, large cichlids. Which species are especially well suited for a community tank with L-numbers will be described later.

Now, we turn towards the topic of the right water for loricariids, the most basic element your fishes need for their well-being.

As you already know, most loricariids that are marketed regularly are endemic to tropical regions and, thus, need water temperatures of 24-26°C. Only species from Paraguay prefer temperatures that are a little bit lower, around 23°C.

Specialised species, like those living in streams (e.g. *Chaetostoma*), do best at room temperature. Regarding water chemistry, one can say that loricariids are really unproblematic: pH between 6 and 7.5 at medium hardness values are accepted.

All Peckoltia *species that are known at the moment, stay relatively small and are thus suited for almost any tank size, like the L 162 shown here.*

But one has to be really careful when it comes to Ancanthicus *and* Pseudacanthicus *species. The L-186 (middle) with its 20 cm length might appear really large, but it is only a young specimen, as you can see when you compare it to the impressive (adult) fish shown in the photo beneath.*

AQUALOG *Special* The most beautiful L-numbers

The suitable tank for Loricariids
The right water

This doesn't mean, of course, that keepers of armoured cats do not have to take care of the tank water - a vast majority of loricariids love well-oxygenated, clean, crystal-clear water. If you are a 'beginner' with the wish to keep the beautiful armoured cats in your first ever aquarium, you should stick to the following procedure: First, you measure the basic values of your tap water. If you don't have the required equipment or test kits, take one liter of fresh tap water to your local dealer who will willingly carry out the measurements for his 'future customer'.

If the resulting pH-value is between 6.5 and 7 and the hardness below 15°DGH, you can use the tap water for most species.

If the values are above these limits, i.e. over pH 8° and 15°DGH, you have two options: Either you decide to do without loricariids and set up a tank with African lake cichlids for which the high values would be perfect, or you buy a water conditioning device, like a reverse-osmosis device that conjures the perfect water (with pH 6.5 and hardness 4°-8°DGH) out of your tap water - the basis for an Amazonas tank.

In any case, we would recommend to read beforehand one of AQUALOG's "Decorative Aquaria" booklets, either "The Amazonas Aquarium", or "The Malawi (or: Tanganyika) Aquarium" in which the setup of such environmental tanks is described exactly and in great detail.

One more 'insider' tip: If you should find out that the calcium carbonate hardness is higher than the total hardness, the reason for this could be a decalcifying device for the water. Water that has run through such a device cannot be used as aquarium water. Please check the water's values before it runs through the decalcifying device and, if possible, use this unconditioned water for your aquarium.

That's about all we want to say about water chemistry. But let us add a few words about the nitrite/nitrate problem. If your tank is well furnished and tended, if your fish don't get too much food, and if you perform regular partial water changes, poisonous nitrite should not be a problem that you (and your fishes!) have to face. Still, nitrite is a highly toxic ammonium product that has a devastating effect on your fishes, even in minute amounts (the absolute top - and already critical - limit is 1 mg/l).

Every now and then, you should carry out a nitrate test. Nitrate is not as harmful as nitrite for your fishes, but in higher doses it can have a negative influence on the growth of young fish and even lead to deformations. Extremely high concentrations would even harm adult specimens. In this respect one should also be aware of the fact that in some regions, the water contains already very high amounts of nitrate when it comes out of the tap so that the legal maximum concentration of 50 mg/l is only just missed.

Another chapter of this book informs you about some symptoms of pollution induced trauma and its immediate treatment. For now, it is probably enough to remember that symptoms like obvious unease, scratching, and increased breathing rates are not necessarily the result of an infection - these symptoms can also be signs of nitrite intoxication.

Chaetostoma *species (here: Ch. sp. from Columbia) like cool, oxygen-rich water.*

Glyptoperichthys, Hypostomus & Co. often live in muddy, slowly flowing and quite warm waters. The fish shown here is probably an adult specimen of L - 083.

Photo of an L-numbers biotope

These two photographs were taken in Brazil, on a boat trip; they both show the same river. Above, you can see a sidearm with cloudy, warm water, the photo to the right gives an impression of a small tributary with crystal-clear, oxygen-rich water. Of course, completely different species live in these two biotopes. Still, we chose to show these two pictures to demonstrate that the collecting place as the only source of information allowing conclusions about the requirements of an aquarium fish is far from being sufficient.

The suitable tank for Loricariids
Setting up the new aquarium:

Furnishing the tank: General tips

Owing to the place where they come from, L-numbers are mostly kept in South America or Amazonas tanks. Here are some basic rules you should stick to:

1. What kind of tank

The tank should not be made of acrylic glass or any other plastic material, because the permanent rasping activities of the catfishes leave these materials unsightly after a while.

2. The tank size

The tank size should always correspond with the size of the fully-grown fishes that live in it. In the captions of every AQUALOG you find the minimum size required by the species in question.

3. Hiding places

Armoured cats need hiding places. The most practicable and also decorative as well as true-to-nature furniture are roots. You can also use plastic tubes, bamboo sticks, clay pots or rock constructions (the rocks and stones may not contain any metal!) - whatever you prefer.

4. The substratum

The substratum must not have any sharp edges or abrasive surfaces, because loricariids always stay in close contact with the bottom where they permanently search for food. Some species even bury themselves in the ground. Of course, a South America tank should never contain a substrate capable of releasing calcium because this would turn the water hard. The best substratum for a loricariid tank is not too fine sand.

5. The medicine

In case, you find that your fishes have contracted some kind of disease, it is important to know that most catfishes do not tolerate any metals or metal compounds in the water. Some medicines, especially those developed for treating oodinium contain copper salts - so please be careful when choosing medicines for disease treatment. For more information about this issue, please read page 35 of this book.

6. The roots

The most important piece of furniture in a loricariid tank are roots, not only because they make an effective and attractive hiding place but they are an important, even vital, part of the cats' diet. You should provide soft wood whose surface the fish can rasp off; slowly, but surely, the fish 'eat' the whole piece of wood, bit after bit. The wood is essential extra food, we will explore this topic in the chapter "feeding" a little further.

7. The filtration system

Like already mentioned, almost all loricariids love crystal-clear water. Eating habits, whether rasping off wood or sticking exclusively to vegetable food, like some species do, might cause a deterioration of the water quality. Thus it is very important to have either a high capacity filtration system in your tank or plan extra partial water changes beside the ones you have to make anyway. It is also very important to clean the filter regularly. Still, if you feel the need to change the filter material, you must never change the whole material but only the wad of floss - all other media are carefully rinsed under a strong stream of water.

You should avoid to replace filter material and change a larger quantity of water at the same time, this would disturb the biological balance in your aquarium completely.

The suitable tank for Loricariids
The right plants

A tank containing only loricariids or in community with other fishes shouldn't be too brightly lit. Usually, armoured cats are crepuscular, that means, only active at dusk and at dawn which plays, of course, an important role in feeding the fish. But - we'll come back to this later. Furnishing the tank with a carefully constructed pile of tangled roots and driftwood isn't just a natural and beautiful way to set up an environmental aquarium, it also provides the 'shade' your catfishes long for. For such a construction, you should always use only one sort of wood. A combination of different types of wood would look unnatural and 'busy'. As you know by now, loricariids are active at dawn and at dusk, they prefer dim or 'half-light'. On the other hand, most water plants need rather a lot of light for proper growth. One solution for this dilemma is to cover the tank surface with floating water plants. Unfortunately, most floating plants do not like it when the water moves too much - but a loricariid tank has to be heavily filtered. Thus, the water is always in motion. Still, there are plants that are indeed suitable for a half-light tank, like, for example, plants from the *Anubias* or *Cryptocoryne* genera. With these plants, you can create a beautiful underwater garden, because they are absolutely undemanding in terms of the quantity of light they need. All *Anubias* species can be used without exception, in *Cryptocoryne* species the exact maintenance details have to be enquired from the dealer. The same is true for Amazon swordplants (*Echinodorus*) which make wonderful solitaire plants wherever free space is available in the tank. As a rule of thumb, one can say that *Echinodorus* species with longer petioles and heart-shaped leaves need more light than the lancelet-like forms. *Cryptocoryne* and *Echino-dorus* species need nutrient-rich substrates to grow in, Anubias species and the ferns described below get all they need directly from the water. Plants do not always have to be planted in the substrate, there are indeed species that do nicely on roots or pieces of driftwood. One of these plants is the Java fern, *Microsorium pteropus*, that attaches its roots to solid wood surfaces. The best way to establish this plants in an aquarium is to tie its roots to a piece of wood until they cling to it by themselves. The attaching process can be accelerated by using soft wood that is recommended for a loricariid tank anyway. If the occasional plant dies because its roots have vanished together with the wood they had grown on you shouldn't be too sad. Make a note under "feeding costs" and buy another one...You can also grow all Anubias species and the African water fern (*Bolbitis heudelotii*) attached to driftwood or roots, thus creating a beautiful aquascape. The usually very popular and useful Java moss *(Vesicularia dubyana)* should be used with caution, because some catfish species have the tendency to get tangled up in this plant. You can give the different tropical water lilies (*Nymphaea*) a try if you have a tank that is not covered. One more extra tip: Sometimes, aquarists are really unfortunate when it comes to planting their tank - no plant whatsoever wants to grow. Some hobbyists then try to establish plants that grow usually out of the water, like *Scindapsus* and *Monstera* species. Seedlings of these plants with well developed roots are given into the tank which looks really nice and also cleanses the water of certain harmful substances.

Here you can see an example for a tank with plants and woods.

But be careful: Some of these plants are poisonous for plant-eating fish and thus absolutely inadequate for being used in ornamental fish tanks. For both, fishes and plants, the installation of a proper working thermostatic heater is most important.

The suitable tank for Loricariids
The right plants

As you can see in the two examples we show here, one can plant a loricariid aquarium beautifully with few aquatic plant species. The photo on the left shows a tank planted with Anubias barteri var. nana, *the picture on the right an aquarium with Java fern.*

Anubias *species are the first and foremost choice for a loricariid tank. To the very left, you can see a large growing species,* A. afzelli, *to the right, the middle-sized* A. barteri var. barteri. *These plants like shade, just like the armoured cats. Planting* Anubias *in the substrate, one has to take care that the rhizome is not covered with sand. For making roots and driftwood look even more attractive, you can use the beautiful Java fern that attaches itself to these kind of substrates. One breeding form is* M. pteropus "Windeløv".

Another, very good choice for your catfish tank is the African fern, Bolbitis heudelotii, *especially if you have catfish species that like water movement - this plant does nicely in water with a rather strong current.*

Whether you use the common hanging type or a submergible one, depends on your personal taste; its primal function is to keep the water temperature at a constant level. In the tropical regions where most of the available catfishes come from the water temperatures stay almost constant, at day and at night. The ideal temperature for almost all cats is between 26° and 28°C. All plants presented here have, besides the characteristics mentioned above, one other, undeniable advantage: they have very hard leaves. This is important because some cats rasp algae off the leaves' surface - in more delicate water plant species, this could lead to whole leaf ending in the cat's stomach.

Community tanks with different loricariid species
Basic rules

The fish's appearance gives some clues about its natural environment and way of living. The impressive L-025 "Scarlett" with its large fins and numerous sharp spines is quite the opposite of an 'aerodynamically' build fish. This leads to the conclusion that in the wild, this fish prefers quiet waters. Interacting with specimens of its own kind, this cat can be quite rough.

Chaetostoma species, like this L-188 from Columbia, are perfectly adapted to living in brooks. Their enormous sucker mouth enables them to hold on tightly to stones or rocks and rasp off algae, even in fast running streams. Among each other, these fish are absolutely peaceful.

Otocinclus species, like this Otocinclus sp. "negros" (no L-number) show a funny behaviour in the aquarium: they attach to water plants in an upside-down fashion. In nature, they live in moderately flowing brooks. Otocinclus species are extreme swarmfish.

Community tanks with different loricariid species
Basic rules

Fortunately, these days the wish to have a tank crowded with colourful fishes from about everywhere is dying out more and more.

Another, fortunate development sees the abandoning of scarcely furnished aquaria with one single large fish.

An aquarium has to be regarded as an important, beautifying part of a harmonic whole. The creation of this harmony starts with the choice of where to place the aquarium and ends with the thought-out furnishing of the tank, thus creating a true-to-nature miniature underwater landscape in your home.

Obviously, only a tank with lively, healthy fishes, swimming merrily around in their artificial biotope will have the fascinating, and, at the same time, soothing effect on the onlooker that is sought by many people who set up an aquarium in their home.

Doctors who have noticed the beneficial effect aquaria can have especially on people under stress more and more recommend the aquarium hobby as therapy to their patients.

Fishes kept in aquaria feel only well if they are kept in community with the right species. The hobbyists has to be really careful with the selection of the different species he/she intends to incorporate with each other.

Today, however, the task of selection is much easier than it used to be, simply because nowadays it is quite simple to get hold of the fishes in question. The pet shops offer a wide range of fish species, so that several different, but still "environmentally correct" communities can be selected.

Basically, you should at least choose only fishes that come from the same continent. Often, such a choice is not only sensible from an aesthetic point of view, it also makes sense in terms of habitat preferences. It goes without saying that one never keeps enemies together in the same tank. Only sadists would do such a thing, but luckily, I have never heard of any hobbyists doing such a cruel thing on purpose. For the fish, living in a sensibly furnished miniature-biotope can be even more "humane" than the rather though life in the wild.

Think, for example, of the Cardinal tetra, *Paracheirodon axelrodi*, that falls prey to birds and larger fishes in huge numbers at the beginning of the dry season.

This "loss" of lives is, of course, compensated when the rainy season begins, with the hatching of millions of young fish. Still, under aquarium conditions, a Cardinal tetra can reach the impressive age of 10 years!

Back to our loricariids: Like in most other fish genera, in these cats not all species can be kept together in the same tank without problems.

Also, one shouldn't keep suckermouth cats only because they make perfect cleaners of your aquarium panes. This habit is certainly a very useful one but the fish do not eat algae to please their owners but because they obviously like to eat them and they are an important part of the cats' diet.

As already mentioned, in the L-numbers, you find small species, like *Peckoltia* with 6-8 cm adult length and large-growing fish like *Pseudorinelepis* with a maximum length of around 40-50 cm.

So, the basic question is time and time again: Can I provide sufficient space? Is my aquarium large enough? Only if you can answer this question with an honest YES you should start selecting the fishes for your tank.

The animals offered at the shops are usually still young, so please take a look into your AQUALOG where the maximum lengths of your future tank residents are indicated.

As a real 'individualist' among the loricariids, *Pseudacanthicus* (L-114) has gained itself quite a reputation. In smaller aquaria where the inhabitants meet quite often L-114 not only terrorises other catfishes but also all other fishes, especially at feeding and when this loricariid feels the need to defend a hiding place or territory.

Other *Pseudacanthicus* species, like, for example, L-079, have turned out to be much more peaceful in this respect.

Community tanks with different loricariid species
Behaviour towards other fishes

If you observe quarrels between males of the genera *Ancistrus*, *Peckoltia* or *Chaetostoma* you don't have to worry about that: if enough space is available, such quarrels might turn up every now and then, but generally the fish will keep out of each other's way.

Some armoured catfish species prefer more of a 'social life', like the fishes of the genus *Baryancistrus* (L-019, L-081, L-177) that are very gregarious especially as young fish. *Cochliodon*, *Hypancistrus* and *Panaque* live relatively peaceful together, even at old age.

Keeping the basic rules in mind, one can indeed keep different loricariid species in a community tank. If you intend to breed your L-number, you have to keep it in a separate species tank anyway, but this topic is treated in more detail in the chapter about breeding.

As the different fish species need different foods, you have to take extra care that these different needs are seen to. The following pages hold detailed instructions how to feed a fish community

In case you don't want to breed your suckermouth cats but set up a community tank with L-numbers and other species and genera, this is possible, too, if all necessary requirements for such a tank are met.

The most basic rule that guarantees a smoothly 'functioning' fish community is to take the proportions of the fishes' sizes to each other into consideration. For example, the small *Ancistrus*, *Cochliodon* and *Rineloricaria* species are the perfect inhabitants of the bottom region in a tank with small South American tetra species which live in the middle and upper regions of a tank.

The larger species like L-011, L-017, L-034, L-047, and L-048, to name some of them, usually do quite well in a tank together with quiet South American cichlids.

Finally, there are the 'giants' among the loricariids, like L-007 (*Leporacanthicus galaxias*), L-014 (*Scobinancistrus aureatus*) and L-025 (*Pseudacanthicus* cf. *serratus*) and others. Presuming you have the space to set up such a huge aquarium required for tending these big animals, there is no reason why you shouldn't keep them in a community tank. As these fishes lack completely a 'social' or friendly behaviour, heavy fights for the best hiding places and territories can break out. But this is the way the rules of nature work - the inferior animal has to retreat.

If the tank is large enough to provide sufficient space for such a retreat, the outcome of the inevitable fights is, most of the time, not too dramatic.

If the inferior species finds an available niche where it can survive despite the presence of a seemingly life-threatening co-inhabitant of the same biotope, it can, just like in nature, exist in this environment.

To cut a long story short: The better you prepare your tank for the size and number of fishes you keep in it, the better you can handle the care of its inhabitants.

An aquarium can be furnished in the best possible way - as long as it is too crowded or inhabited with species that have to fight for their lives all the time, it cannot hold any attraction for its owner.

Besides - fights, especially among larger fishes, leave your aquarium furniture in a state of absolute mess; all the trouble you took to set up a nature-like environment for your pets will end up in chaos and the furniture will look like having suffered from a heavy earthquake.

Unfortunately, there is not much information available about the large loricariid species (compared to what we know about the smaller ones), so that it is impossible to say "Take this one and this one, and everything is going to be all right...".

This situation owes to the fact that the large L-numbers are only known for a very short while in the hobby so that the opportunities to study their behaviour have been quite rare. Also, one can basically say that these big fish prefer a rather 'individualistic' lifestyle - one can never be sure how they will react in certain situations. So - if you want to keep several of such large specimens or several large species in a community tank you'd better be prepared for an emergency situation... just in case your community tank turns out to be a war

Community tanks with different loricariid species
Behaviour towards other fishes

L-114 is an absolutely smashing, but also very quarrelsome loricariid.

zone. If you incorporate large loricariids in a tank with cichlids you can expect territory fights and other quarrels. But this kind of trouble will cease after some days - well, if there's enough room for all inhabitants.

In case the fights continue, you have to separate the fishes from each other, and keep each species in its own tank. After a while, you can try again to set up a harmonious community, sometimes it works at the second go. If you don't have a spare tank for the separation you could ask other hobbyists you know or an aquarist society (if you are a member) for 'asylum' for your fish.

If you don't have any of these 'emergency accommodations' at your disposal you should give up the idea of such a community tank experiment. Instead, we recommend to enjoy these wonderful animals in either your AQUALOG book or in the display tanks of a public aquarium.

It is better to stick to the small, unproblematic, (and also beautiful!) species than to unnecessarily risk the life of an animal.

Still, in small species the same fights for food or territories can break out, although they are often not considered as bad as in large fish; somehow, the aggressive behaviour is - when displayed by small fishes - regarded as playing, which, of course, it is not.

You have to have a keen eye on all your tank inhabitants. If weaker fish are constantly driven away from food or forced like sheep into some corner of the aquarium, you have to intervene and separate the opponents from each other.

Luckily, in smaller species this problem occurs quite rarely, probably owing to the fact that small fishes very often are kept in more spacious tanks than large ones. Nevertheless, disharmony in a community tank can turn into quite a harmonic status quo, for example, after breeding pairs have formed or the hierarchy has been settled.

If no settling of affairs whatsoever has come about after a while, you have to proceed as described above, in the section about the large cats.

Community tanks with different loricariid species
Behaviour towards other fishes

The only difference is the required tank size. Such an 'emergency tank' should be in any aquarist's home, anyway, as a quarantine aquarium for new fishes.

New fishes intended for a community tank should always be put under quarantine for a couple of weeks before they are incorporated into the established tank.

A separate tank is also very useful when some kind of disease breaks out in your community tank. Most of the 'common' diseases can be successfully treated with the usual medicines.

Let's turn once more to the topic of establishing a fish community with L-numbers in captivity. For disease prevention, it is most important that the tank inhabitants can live as stress-free as possible. There is an easy rule of thumb: the less stress, the less diseases.Like in all other living organisms, diseases are latent and can break out any time when proper care is neglected or stressful situations get out of control.

It is beyond the scope of this advisory book to list every possible combination of fishes that would make a beautiful and environmentally sensible community tank.

Still, we can offer you some suggestions, based on experience and practical knowledge acquired in long years of fishkeeping.

Unfortunately, most armoured catfishes react extremely negative to many medicines.

Therefore, it is practical to have a separate tank at disposal where the sensitive fishes can be tended during the phase of disease treatment.

We'll come back to this in the chapter "Disease Treatment".

There might turn up problems, though, like in any other community or situation where living creatures are involved; if this is the case, please proceed like described in the section above.

Some peaceful large L-numbers, like this Glyptoperichthys gibbiceps, can be unproblematically combined with smaller fishes. As you can see in this picture, the small Apistogramma is hardly impressed by the cat's behaviour.

Community tanks with different loricariid species
Choosing fishes for your tank - suggestions

Tetras are the most important fishes in South America tanks. Still, not all tetras are alike. There are those that live in the upper regions of an aquarium: hatchetfishes (*Carnegiella, Gasteropelecus, Thoracocharax*) can be kept together with almost any L-number without any problems, simply because the fishes have completely different swimming habits. *Copella* and *Thayeria* species also live surface-oriented. All these fishes can be recommended as tank comrades of loricariids without any reservations.

The same is true for fast swimming swarm fishes that live in the middle regions of the water, like the characoids *Aphyocharax* and *Prionobrama*.

Species living bottom-oriented, like *Nematobrycon* and *Impatichthys* species, as well as the popular neon tetras (*Paracheirodon*) and the numerous small *Hyphressobrycon* species can only live harmoniously together with small, peaceful L-numbers.

Robust, sometimes quarrelsome fishes from the headstander or *Leporinus* family should be kept only with larger cats which are just as robust.

Such tetras are, for example: *Abramites, Anostomus, Leporinus* and *Semaprochilodus*.
As these fishes need quite a lot of space, according to their size, the future owner of one or several large tetra species will have to inform him/herself about more detailed care instructions at the pet shop or other specialist literature.

Just let me tell you this one thing: If you want to keep these tetras, you should always have them in swarms of 8 to 10 fish.
If you don't, these species are especially susceptible to abnormal social behaviour that is shown through aggressiveness towards members of its own species as well as other fishes.

Wonderful tankmates for large community tanks are the different 'silver dollars'. They are an attractive contrast to the more quiet suckermouth cats.

But before you buy 'silver dollars' for your community tank, you should read specialist literature on these fishes, because their adult sizes differ considerably and the young specimens of these species look very much alike and are hard to distinguish from each other.

There are species with 10-12 cm maximum size that are still suitable for a one meter tank.

Others can attain up to 50 cm length! Although these maximum sizes are rarely reached under aquarium conditions, you should refrain from buying large growing species if you cannot provide the space they need.

Now, let's turn to the second, large group of South American aquarium fishes, the cichlids (Cichlidae). Keeping loricariids together with discus (*Symphysodon*) is not unproblematic, as many L-numbers do not tolerate the high water temperatures the discus need.

Other suckermouth cats have the (for the discus!) unpleasant habit to search the large surface of the discus body for food to rasp off.
Experience showed that the peaceful *Hypostomus, Glyptoperichthys* and *Ancistrus* species do quite well in discus tanks. Unfortunately, we cannot give you any reliable tips regarding this special fish community, you have to find out for yourself if it works or not.

Anyway, always have an 'emergency tank' prepared, just in case the combination discus/loricariid doesn't work at all.

Many of the small and peaceful suckermouth cats are well suited to be kept together with angelfish, especially because the shy angels like the subdued atmosphere of a loricariid tank.

The larger cichlids like oscars (*Astronotus ocellatus*), the cichlids from the *Cichlasoma* kind, and *Aequidens* or *Crenicichla* species should be tank comrades of robust, large L-numbers.

The aquarium furniture should provide many hiding places where the cats can hide during the period when the cichlids care for their offspring.

Community tanks with different loricariid species
Choosing fishes for your tank-suggestions

Also, one should not forget that even primarily algae-eating catfishes cannot resist a tasty meal like the spawning of a cichlid if it is within their reach.

So, if you want to breed your cichlids in a tank with loricariids, always keep a light on at night so that the cichlid parents can see and protect their eggs.

If you want to keep loricariids with eartheaters like *Geophagus* or *Satanoperca* species, you need a little experience in fishkeeping, because these large, bottom-oriented cichlids will inevitably compete with the cats for room and food.

Other catfishes, like plated cats (*Corydoras*) are suitable tank inmates for aquaria with small L-numbers. Large cats, like *Sorubim lima*, the Shovelnose catfish, are the ideal company for larger loricariids.

Generally, one can say that predatory fishes (like the mentioned Shovelnose cat and predatory tetras) do fine in tanks with L-numbers, if the cats are large enough to be not considered a tasty meal. If they were, this could be fatal for both sides as armoured catfishes have a very effective passive defence system, as you know from the introduction. Predators can choke to death on such a supposed delicacy.

Species that are usually kept in species tanks like killis (for example, *Rivulus* species) or Leaf fish (*Monocirrhus polyacanthus*) can be combined with small suckermouth cats without any problems as these fishes do not influence each other's way of living; quite the opposite: the little 'cleaners' keep the aquarium climate healthy.

Livebearers like platies, guppies, swordtails etc. can only be kept with L-number species that tolerate the intense light levels livebearers need. Another point is that most livebearing species need hard, alkaline water which is the opposite of the water conditions the catfishes require.

Well, these more general instructions are about all we can give at the moment. It is a simple fact that no tip whatsoever can replace your own observations and experiences.

It is always a good idea to take notes of everything you observe when you start with the wonderful aquarium hobby - and it is just as good an idea to publish your observations in one way or the other so that all aquarists can share what you have experienced. Because one thing's for sure: The constant exchange of knowledge about our beloved ornamental fishes will add to joy they bring into our lives.

When it comes to feeding loricariids, there are several important factors that have to be considered. In this chapter, we want to go into these factors and offer you advice from our own experience.

First, let's see what kind of food armoured catfishes live on in nature. The overwhelming majority of armoured cats do not tolerate temperatures below 16° C, therefore you have to make sure that they do not cool down too much when they transported in a plastic bag.

In all probability, this inseparable connection of armoured cats and tropical temperatures derives from the specialised feeding habits of these fishes. They live substrate-oriented and with their suctorial mouth with genus-dependent teeth they rasp their food (algae, small crustaceans or insect larvae) from the substrate they are attached to: tree trunks, rocks or sandy bottoms. This special mouth gave the cats their other, commonly used name: suckermouth cats.

Often, the food the fish take in is hard to digest and nutrient-deficient. In order to be able to get the nutrients from the sometimes indigestible food, the fishes have special, symbiotic living bacteria in their (very long) intestines. Still, it is believed that only in warm water enough organisms live to provide sufficient nourishment.

Maintenance:
The right foods

Looking into the fish's mouth tells you a lot about the eating habits of a species. L-051 (top, left) has only one row of fine teeth on each side of its mouth. Fishes with such a mouth eat algae or plants growing on any kind of substrate.

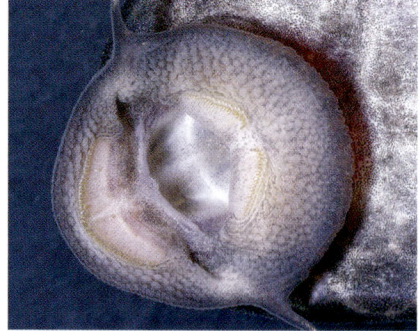

In L-163 (centre, right) each jaw has two rows of strong teeth. This is a perfect rasping instrument that can even crush wood without problems.

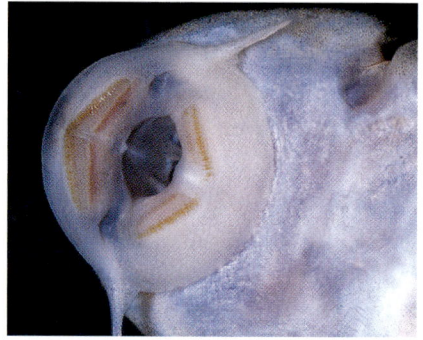

L-200 (centre, right) is an example for the next step: two rows of teeth on each jawbone, but these are long enough to work also as a grasping instrument. This fish will eat plants as well as rotting flesh: it is an omnivore.

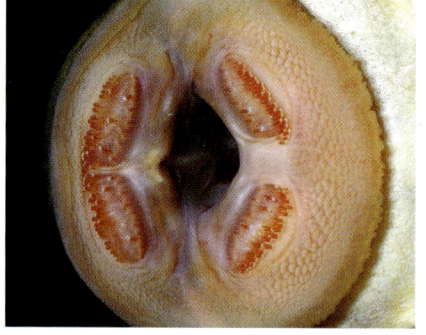

The typical mouth of a carnivorous fish can be seen in L-160: Interlocking, long and strong teeth are no longer useful for rasping - these teeth are made for biting off.

But this explanation still falls short when it comes to the question why, then, most carnivorous suckermouth cat species do not live in colder areas. The answer is quite easy: in the carnivorous species, you have either the older, "original" forms or secondary specialised ones - the evolution of suckermouth cats develops towards a purely substrate-oriented animal.

So far, this development has been really successful; the ecological niche the suckermouth cats have occupied foodwise sees no competition - and this is also the reason for the incredible multitude of Loricariidae species. -

Maintenance: Feeding

As mentioned before, not all L-numbers are 'vegetarians'. To make it easier for you to get basic information about the feeding habits of each loricariid we added a symbol to each picture that's included in the poster:

(☉) = herbivorous / vegetable food
(☺) = carnivorous / meat food
(◇) = limnivorous/ food growing on surfaces
(☺) = omnivorous / mixed food

As the different types of teeth already indicate, there are loricariids that have specialised in certain foods or ways of eating. Still, there is no detailed description available which is probably due to the fact that most L-numbers have been discovered only a short while ago. Ethology (like any scientific study) requires intense research over a long period in order to be exact and meaningful. Observations and examinations have to be carried out not only of aquarium fish but also of specimens in their natural environment, which poses sometimes serious trouble on the researcher.

Nevertheless, one can quite securely rely on the experience of importers, breeders and aquarists as many loricariid species have been successfully bred in captivity. And such a success is only possible if the fishes are correctly fed.

The genera *Pseudacanthicus, Scobinancistrus, Hypancistrus* and *Leporacanthicus* detest vegetables, they prefer meaty material. These cats should be fed frozen food like, for example, insect (fly) larvae, *Artemia* nauplii, and beef heart - a whole range of such foods are offered in the pet shops. They also take frozen or fresh fish - but be careful: Do not feed trout, for some unknown reason, some species cannot digest it.

The vegetarians among the suckermouth cats, *Peckoltia, Ancistrus, Cochliodon, Baryancistrus, Hypostomus, Glyptoperichthys, Liposarcus* and *Panaque,* prefer a diet consisting principally, not exclusively, of vegetable food.

These fishes will also eat meaty material when it is offered but a carnivore diet is not sufficient for them. Without feeding additional vegetable food, your plants would very soon suffer from being considered part of the daily diet...

Feeding fruit and vegetables is not only cheaper than buying new plants; a tank with healthy and 'complete' plants also looks much better and keeps the delicate biological balance of your tank intact. A wide range of fresh vegetable foods is readily available.

This L-014 shows beautifully its protrusible mouth.

Feed, for example, thinly sliced fresh cucumber or zucchini (without the skin), potato slices, mashed peas, carrots, blanched lettuce (rinse well!), spinach or Brussels sprouts (both either fresh, boiled or frozen).

You can also give some fruits, like bananas (of course, without skin!), mango, papaya or mashed coconut. All these tasty fruits grow in the countries where your fishes come from and are part of the natural foods available to the animals.

Maintenance:
Food and roots

Do not feed apples or citrus fruits because they contain acids that can turn the aquarium water cloudy.

You may wonder why we make such a long, detailed list of vegetables and fruits.

Well, like I already said, the L-numbers are very sensitive food-specialists.

They all have different tastes and you have to find out for yourself which food is preferred by the cats you own.

This is somewhat time-consuming but also interesting and makes feeding maybe a little bit more exciting than it usually is.

Another important factor in loricariid feeding is wood. It has been claimed that some loricariid species live exclusively on rasping wood - at least, some *Panaque* specimens that were examined had wood as the only filling in their intestines.

I know from my personal experience that almost all L-numbers need wood as part of their diet.

For the keeper, this requirement is no problem at all, because an environmental tank includes roots and driftwood as part of the furniture anyway.

Probably the best way is to offer hardwood roots that resist the cats' rasping for a very long time and add, depending on the tank size and number of loricariids kept in it, a number of softwood roots as well.

These latter roots are eaten up after some time and are then replaced by new ones.

One more tip: When setting up a new tank as well as when adding roots as part of the food you should put them shortly into boiling water so that any germs that might be there are killed off.

Also, the soaked roots are heavier and stay at the bottom of the tank.

Last, but not least, they do not release as much of the tannin that gives your aquarium water an unsightly brown colour if you don't waterlog the roots beforehand. Of course, you may only use commercially available moor roots. Never use roots you have freshly collected by yourself, except you found them in a moor.

Fresh roots start to rot when you put them into water and this has a disastrous effect on your aquarium.

Now we have talked so much about the specialised feeding habits of loricariids that you might think it is difficult to feed them correctly.

But don't be scared: Nowadays, such a wide range of custom-made foods, either frozen, tablet or flake, is available that it is no problem at all to keep and feed L-numbers.

And the try-out of the personal vegetable and fruit taste of your catfishes might even turn out to be quite amusing.

When feeding, it is also important to remember that most loricariids are nocturnal animals, that means, they are active at night and rest during the day.

Accordingly, they are out searching for food when it's dark.

In the aquarium, and especially in aquaria set up with dim lighting, they usually adapt quite quickly to your feeding rhythm and come out of their hiding places to eat during the day, too.

Now, the last, but also very important thing you have to think about is the fact that the bottom-dwelling cats live on the food that is left by the other tank inhabitants and then sinks to the bottom.

I cannot recommend to feed simply more food; the additional food and left-overs would spread in the whole tank and be sucked in by the filter, thus unnecessarily polluting the water.

You could turn off the filter during feeding, but this is quite a dangerous solution because it is obvious that forgetting to switched it on afterwards could have a rather hazardous effect...

Maintenance:
Food and roots

There' one rule of thumb in feeding aquarium fish: Always feed as much as they greedily eat within a few minutes. You should stick to this rule, because most fishes living in captivity are fed too much and are much too fat.

Considering the way of how to feed the fishes swimming in the upper and middle regions of the tank correctly, one has, in order to supply the bottom-dwellers with food, too, feed more often than usual. Of course, if you feed more frequently, you feed less at once; I also recommend to offer, at the same time, flake, tablet and frozen food, because even the most greedy fish cannot eat everything at once. This way, one or the other food can sink to the bottom-dwelling species.

Besides this 'technique', there is another, more elegant solution of the problem: You could buy one of the automatic feeders offered in the pet shops - you probably need one anyway, the next time you want to go on holiday. Fill it with the given foods and set the clock to release the food at night-time when no other fish is on the search for food. With such an machine, the loricariids will definitely get their share.

By the way: When it comes to feeding, the vegetarians among the catfishes have an undeniable advantage over their meat eating brothers: the raw cucumber or potato slices are only enjoyed by those cats that really like vegetables!

Up to this point, we talked about several aspects of keeping the interesting and beautiful loricariids, where they come from, the best care, the optimal foods. Which leaves another basic question open: Can these fishes contract diseases?

Well, of course: Loricariids, like all other animals, can fall ill. Most diseases are latent in every organism and can break out when the organism is distressed in some way, like, for example, through disturbance of the usual way of living or continuous stress.

Or through negative changes of the animal's environment - which leads us, the aquarists, to the water our fishes live in.

Almost all fish diseases that crossed my way in the many years I spend as an aquarium enthusiast were - as far as they were diagnosed and their causes analysed - the result of a deterioration in water quality.

Therefore, it is most important to regularly carry out all necessary steps of maintenance. One of these measures is the regular partial water change which shouldn't be done only when the fishes already display signs of distress.

Just imagine, the pH of the water had steadily fallen to pH 4 (because you forgot the water change). If you change water at this stage, using water with an pH of 7.8, you can probably picture the shock the fishes will get from this procedure.

A similar thing would happen if you changed the whole filtration system at once. So, it is important to clean and exchange parts of the filter regularly, but never all at once and never at the same time you change the water.

The frequent measurement of the water parameters should become maintenance routine for any aquarist. Controlling the water's hardness (DH) is not necessary if you use water from the same source all the time; theoretically, the hardness never changes and can thus not be responsible for any disease breakouts. The most important value in a tank is the nitrite level.

Nitrite shouldn't be measurable at all in your aquarium - a level as low as 0.5 mg/l can already be harmful for your fishes.

If you find dissolved nitrite concentrations at or even above this level, you have to make an immediate partial water change.

If such a high nitrite level is measured although the necessary maintenance programme has been carried through regularly, one has to search for the reasons for this dangerous intoxication of the aquarium water.

Very often, it is caused by excessive feeding, rotting plants, an overstocked tank or an ineffective because too weak filtration system.

Maintenance:
The settling in period

Measuring the pH value is just as important; in loricariid tanks, it should be between 6.5 and 7.5. In unstable water (DH below 10), the pH can easily decline and also decomposing food left-overs might be the reason for an unhealthy turn of the water's pH into the acidic range.

The captions' symbol text includes the pH recommended by us for each fish presented on the poster. Please note that we purposefully did not take the pH of the water the fish comes from but the pH we experienced to be correct for keeping the fish in an aquarium.

Anyway, you shouldn't miss the opportunity to ask the dealer or breeder you buy your fishes from at which pH value he keeps them. Especially the value of a breeding tank can differ considerably from the pH we recommend for the 'usual' keeping, so that the move from the breeder's tank into your aquarium at home can be quite a shock for the animals.

Such a possible discrepancy of the breeder's or dealer's water and the water in your tank can be evened out if you slowly accustom the fish to the tank water through continuously adding more and more tank water to the water you transported the fish in.

Here you can see a new fantastic loricariid without a number.

When setting up a new tank, it should be furnished with equipment and water at least one week before you introduce the fishes: the tank has to be 'run in', just like a car engine. Condition the tank's biology with, for example, "amtra clean starter" and check if all the equipment works properly. Before you introduce the fishes, control the temperature and the water parameters. I also recommend to add a water conditioner like "amtra care". After the 'running in period' of - at least! - one week you can add the fishes to the aquarium:

Float the bag in which you transported them in the tank for about half an hour so that the difference between the temperatures of transport water and aquarium water can adjust.

Then you open the bag and start to slowly change bag water for tank water. For this, you allow some tank water to run into the bag; from time to time, you pour a little water out, thus mixing the water gradually.

During this process you have to keep an eye on how much air the bag contains - if the bag collapses the fishes might suffocate. If the water in the bag is mostly that from the tank, you can release the fishes into the aquarium.

If you only want to add new fishes to an already established tank, you can basically proceed just in the same way. But - "prevention is better than cure".

This is why I recommend to set up a quarantine tank that should be equipped with all the necessary devices (filter, heater etc.), although such a tank can, of course, do without any 'beautifying' furniture.

Introduce the new fishes to this quarantine tank just as described above. Beforehand, you should treat the water with "amtra care".

This is important because such a treatment protects the fish's mucous membranes, helps to heal any wounds inflicted through catching and also absorbs some of the harmful substances that might be dissolved in the water. Keep the new fishes two or three weeks under close observation. If they have any diseases, these will become apparent during the quarantine period. In quarantine, you can treat and heal diseases without straining other fishes and, first and foremost, you prevent all other inhabitants of your community tank from contracting a disease.

During the quarantine period you make a small daily water change: Remove some water from the quarantine tank and fill the tank up with the same amount of water from the main tank.

If all new fishes are healthy (clear signs are good appetite and lively swimming), you can introduce them to their future home.

Maintenance:
The partial water change

Well, what else can be done in order to prevent your pets from falling ill? You guessed right: the regular partial water change is the basis for a healthy aquarium.

But - if it has to be carried out so often, how can it be done without a lot of trouble?

And how can it be done without making a mess (and without, consequently, having a dispute with the wife about the almost unavoidable water stains)?

The experiences of an import/export installation or wholesaler like Aquarium Glaser with over 3000 tanks are hardly useful in this context, because most aquarists have their tanks in their living room, or at least in rooms with carpets or wooden floors.

Fortunately, we know many enthusiasts and one of our friends had the perfect solution to this problem at hand. (Maybe I should mention that his wife is very pernickety when it comes to her precious carpet, so he was forced to develop a technique that works a hundred percent with - out any mess!). This friend of mine proceeds as follows: Before he starts with the water change, he places an exactly fitting piece of an old, thick carpet around his 120 cm tank.

This 'carpet cover' prevents the expensive carpet from any damage even if some water is spilled.

He doesn't make a 25%-30% water change every one to two months, like it is recommended in most specialist literature, but a weekly 10% change.

This is not only sensible but also very close to natural conditions; in the three years he stuck to this 'high frequency' water change, he never had any diseases nor deterioration of water quality in his aquarium, although it is quite densely stocked.

And the water is always crystal-clear and the aquarium as decorative as can be. Another advantage of this smaller water change is the fact that it is less time-consuming: my friend needs as little as 15 minutes to carry it out.

When removing such a small amount of water from a tank, one even can do without expensive devices:

In maintaining young fish, like this baby L-241, the partial water change is especially important; it's the basis for healthy, large growing fish.

It is just as important in tanks with really large fishes (like this L-185) as such big specimens have a very active metabolism.

Just take off the tank top, siphon water and mulm and replace the removed water with a watering can (without the spray attachment) by fresh water conditioned with "amtra care".

When he changes or cleans the filter, he proceeds just as skilful: The filter is disconnected, taken to the toilet, rinsed or changed, connected - that's it!

Even his sceptical wife is now a dedicated friend of the hobby and she told us: "We'll soon set up another, bigger aquarium with African Cichlids, and the TV set will be moved to another room, because it is much more interesting and relaxing to watch the aquarium life than the television programme!"

Well, the number of female enthusiasts rises by the hour - needless to say that the story I told you works the other way round as well...

Maintenace:
Disease treatments

There is one more thing I think is important when it comes to disease prevention:

You should always buy young, "medium" sized fishes. This means, the animals should be not too young, but not fully grown either.

It is like in planting a tree in your garden: If it is too small, it might have problems with taking root, if it is too old, it hardly has a chance. And in fishes that have to adapt to a new environment, it is just the same.

Of course, you should never buy fishes that are obviously ill. But we can hardly imagine that any responsible dealer or breeder deliberately sells ill fishes - for the overwhelming majority of today's dealers it is most important to have customers that are satisfied and buy regularly at the same shop.

Our tip: If you are satisfied with a certain dealer, continue to be his customer and do not buy fishes from different shops, only because they are cheaper or a special offer.

You save yourself and your fishes a lot of trouble and unnecessary stress.

A competent dealer will always be there for his customer to give advice, as good as he can. A trustworthy dealer can save you more money than a dubious special offer!

It is possible, though, that latent diseases break out because the fishes are distressed through transport or the catching procedure in the shop.

As long as the disease is only latent and not apparent, the dealer or breeder cannot cure it, so it is, of course, not his fault if a seemingly healthy fish falls ill shortly after the purchase.

If you put all new fishes in quarantine before you introduce them to an established fish community you can be sure to be safe from any unpleasant surprises.

I want to stress the recommendation never to buy fishes from special offers. It is very important to check whether extremely low prices can be trusted or not, when it comes to aquarium equipment.

But in live fishes - like in any other animal - you should never try to save money. Good quality, healthy fishes do have their price - and that rightly so.

Now, we tried to give you advice for disease prevention - but what to do when the fishes contract a disease, despite all efforts? Well, things like that happen, even doctors are ill very now and then.

It is essential that you are able to ascertain yourself quickly of the tank inhabitants condition so that you can detect and diagnose diseases at the earliest possible stage.

Only the immediate correct treatment of a given disease can guarantee that ill fishes recover fully.

It should become daily routine to check your fish's health. The best opportunity for the daily check-up is, of course, feeding time. Healthy fishes respond quickly to food in the tank and eat greedily.

If your fishes suddenly display a diminished appetite, lose weight, are apathetic and lethargic, rub themselves or show strained breathing, you know that there's trouble ahead.

Basically, there is no miracle cure for diseases. As this is an advisory handbook for L-numbers, it would be far beyond its scope to discuss all known fish diseases and their treatment. We warmly recommend to buy real specialist books, like, for example, by UNTERGASSER, BASSLEER or REICHENBACH-KLINKE; we listed all books we consider useful in the back of this Special.

A well-tended species tank with L-numbers is almost always free from any serious diseases. If diseases do appear in such a tank, they are mostly fungal skin infections of injuries resulting from being netted in the pet shop or fights among the tank inhabitants. Usually, such infections heal by themselves and eventually the fungus will drop off. If not, treat the isolated fish as follows: Raise the water temperature by 3-4°C and add some salt (1 spoon to 10 l water). You can either use the usual cooking salt or sea salt used for marine tanks. In this conditioned water the fungal infection should heal completely.

Maintenance:
Disease treatments

If a fish has a healthy appetite but loses weight nevertheless, this could be a sign for worms. Put the fish into the quarantine tank, so that the other inhabitants are not infected via the ill fish's excrement. Worm infections can be treated with medicines containing either trichlorphone or flubendazole. Both substances are extremely poisonous! These drugs are only available in pharmacies with a vet's prescription. Be careful to use only the amounts recommended in the manufacturer's instruction, or even less.

Unfortunately, loricariids often react negatively to these kinds of drugs, so please be extremely careful when using them. Also, there are the many different kinds of parasites that can infest our fishes. Parasites are rare in loricariids, but never say never...

There is, first of all, the well-known Ichthyophthirius, the White Spot Disease or ich; it can be seen and diagnosed quite well with the naked eye. It might be already too late if you see a fish that's completely covered with white spots. The first sign of a possible infection with ich is a fish that rubs itself against stones or roots or any other objects in the tank.

A soon as you discover white spots where white spots do not belong, you have to react immediately. With the beginning of any suspicious rubbing you raise the water temperature by 2°-3° C for about two weeks which solves the problem - hopefully. If it doesn't, you need to use a drug; ask your pet shop dealer for advice! Please note: catfishes usually tolerate only reduced amounts of the recommended dosis, so please watch your fishes carefully for any signs of distress while treating them.

If you have a species tank, you have to treat the whole aquarium, because the young, invisible parasites that swim freely in the tank have to be killed as well as the adult parasites in the fish's body. If you have catfishes in a community tank you have to put them into a separate tank and treat them there with the reduced dose. The main tank is then treated with the recommended dose as usual.

There is another parasite that is more than unpleasant: Oodinium or Velvet or rust disease. Infection with this probably worst plague of freshwater aquaria manifests itself in form of spots, that are much smaller then in ich, and also not white but more grey or velvet: In the final stage of the disease, the fish is covered with a greyish dust. As the spots are so small, the infection can hardly be seen at an early stage. Sign for an infection with oodinium are diminished appetite, loss of vitality and the already mentioned rubbing of the irritated fish that wants to get rid of the parasites. Treatment is the same as in ich, but you should add high-quality sea salt to the water (2 spoons to 10 l water). Please note: Some catfishes do not tolerate the copper sulphate contained in "amtra medic 1" - unfortunately, there is no effective drug without copper. If you detect signs of intolerance in your catfishes, you have to make an immediate partial water change.

The reason why we recommend "amtra" products is that these products have been developed for professional use and that we have reliable data on how effective they are and how they work. Instead of the recommended treatment you can also try an intensive cure: A very short bath in a high concentration solution. But this treatment should only be tried by experienced aquarists. Anyway, if you are not quite sure how to proceed, ask your vet.

Unfortunately, there aren't too many veterinarians who are fish specialists. If the treatment of your ill fishes expects too much from your vet, maybe an aquarist you know or the breeder you bought your fishes from can help. Very often, especially breeders are very experienced when it comes to diseases and their treatments. Tips about disease prevention and treatment can also be found in the only newspaper for aquarists, the AQUALOGnews that is available to you for free every six weeks at good pet shops.

In the rubric "The Fish Doctor", Dr. Markus Biffar, a veterinarian and fish specialist, answers questions and offers advice on diagnosing and treating fish diseases. If you have a community tank, and all other fishes except the loricariids are affected, you have to isolate them before you treat the main tank with the recommended medicine. Treat the loricariids prophylactically in the isolation tank with raised water temperature and added salt - they could be infected although they look healthy!

Maintenance:
Disease treatments

The specimen of L-174 (top left) with a fungal infection of a wound at the left pectoral fin spine. Good care restores the fish's health but the severely damaged fin will never again grow to its original size.

Smaller defects, like the one of the dorsal fin of the L-239 (centre right) heal usually all by themselves. White growths on the body (like on the head of this L-100 Ancistrus) need intensive care. Such bacterial infections are very dangerous for the animals.

Albino coloration like in the L-003 (top photo: normal coloration, below the albino form) is not a harmful disease but a genetically caused colour form.

Breeding:
Basic requirements

If you ever see an armoured catfish that is totally emaciated, with a turned inwards belly and the eyes sunken almost invisibly into the sockets, this poor creature has either starved for a considerable time, been offered the wrong foods or is heavily infested with parasites/worms. When a fish is in such a miserable physical state, it is probably too late for any treatment because it is no longer able to eat. And if treatment and healing is impossible is it the best thing to destroy the ill fish.

As a conclusion, it can be said that loricariids are hardy, robust fishes that are rarely ill when being kept in a large enough, carefully furnished tank and regular maintenance. A secure sign for fishes that feel really well in their artificial biotope is, of course, reproduction.

There are some really interesting breeding reports about several L-numbers, but there are species whose reproduction patterns are still a complete mystery. This is quite understandable because these fishes have been present in the hobby only for short while and in some species it takes very long to find out how the different sexes of a given species look like and under which conditions they spawn and breed.

The differentiation of the sexes is especially difficult. In some *Ancistrus* species, like, for example, L-071, it is really easy: males have a clearly visible 'moustache', females only very short barbels. In other species, the sexes are hardly distinguishable by any outer characteristics, like L-137: male and female look almost exactly the same.

In many species, males and females can be distinguished when looked at from above. Very often, males have a broader head, whereas females are somewhat shorter and more sturdy. In *Ancistrus* species, the males have very characteristic interopercular odontodes, particularly salient on the sides of the snout (almost looking like whiskers). In L-002, the females develop a thick 'field' of spines on the caudal peduncle, in L-093 (and other *Aphanotorulus*) the males have many small spikes on the upper side of the caudal peduncle. In many genera with flattened bodies (like *Lasiancistrus* or L-238) the males have more and longer spines on the first ray of the pelvic fin. Whether L-018 and L-081 are two different species or only male and female of the same species, has not been decided yet. There are indeed species in which the patterns of male and female are completely different, like in the many Peckoltia species. In all catfishes, there is one characteristic that is the same: the females are clearly more ample than their males. But all the characteristics listed here are very general; for more detailed information, please read IsBRÜCKER's work (1994).

It is not easy to breed armoured catfishes; they do not reproduce all by themselves, like guppies, platies or other breeding forms of livebearers. If your loricariids should spawn and breed in the aquarium in your living room without having been especially conditioned for it, this is certainly a pure coincidence, but also a sign for very thoughtful care and maintenance - the owner of fry resulting from such a spontaneous reproduction can be really proud of himself.

Successful and repeated breeding can only be expected when the loricariid keeper is an experienced enthusiast who indulges himself in fish keeping and has all the required equipment for "professional" breeding.

Still, every successful breeder has started once, and this is why we will try to give you a general insight into the most important basic requirements for loricariid reproduction. This can be by no means a professional instruction on how to breed catfishes. For more detailed information you should buy specialist literature, and collect articles from aquarium magazines or the newspaper AQUALOGnews.

Our tip: Contact experienced breeders, or join a local aquarist society - its members can probably offer advice on one or the other problem.

Then simply start with your breeding efforts; you will see very soon if you have a knack for it and how well you can cope with negative experiences.

To build up a stock, it is not enough to have the usual home aquarium in the living room. You should at least have a suitable space like a garage or cellar where you can set up some tanks and start with the first cautious steps of being a catfish breeder.

Breeding:
Tips and breeding reports

Of course, we do not know whether you are interested in breeding your loricariids at all, but if you do, we wish you the best. One thing is for sure: If you ever succeed in breeding your own catfish-offspring, you will recognise all facets of the wonderful hobby and become a real enthusiast.

Basically, it seems as though the simulation of the beginning of the rainy season is one of the key factors that trigger off successful reproduction.

As soon as the rounded belly of the spawning-ready female can be clearly seen, the rainy season "begins".

First, the water temperature is lowered, step by step, to about 20°C. Feeding is continued as usual, but no water changes are made.

It is important to control the nitrates during this time, the maximum level of 100-120 mg/l must not be passed.

During this 'bad weather period', the female should 'ripen' and develop a swollen, roefilled belly. If your fish start to 'wag' their pelvic and pectoral fins instead, this is a clear sign that the fish do not tolerate the simulated, 'cold' weather; in this case, the breeding attempt has to be stopped immediately!

When things run smoothly, and the female develops roe, the actual spawning is induced. You have to make a daily water change of 60-80 percent of the tank contents, using soft, slightly acidic water.

At the same time, the temperature is raised to 26-28°C and the lighting period is extended to more than 12 hours a day.

After two weeks, the volume of the water change can be reduced to the normal level.

Suitable spawning sites have to be provided. Which one is preferred by your catfish, you have to find out for yourself.

I recommend to offer them all sorts of caves, so that the fish can choose the spawning site they like best. Some species need to dig holes or burrows in the sand or remove sand from a cave in order to continue with the spawning. If the pair harmonises, you have now the conditions to spawn them successfully; sometimes, they already spawn during the 'big water change' phase.

This may sound not too difficult, but in most loricariid species it is really hard to induce a first, initial spawning. But as soon as a pair is established and well-adjusted, they breed quite easily.

Like I already mentioned, it would be beyond the scope of this book to give detailed descriptions and reports of successful breeding attempts of loricariids.

Still, one can generally say that the small species, like *Chaetostoma*, *Hypancistrus* and *Peckoltia*, breeding is more often successful than in others. In *Hypostomus* and other large growing species, breeding is far more difficult and hardly ever successful.

One reason for this could be the often too small tanks these large fishes are kept in.

There are reports about spawning of

L - 080 (*Peckoltia* sp.),
L - 046 (*Hypancistrus zebra*),
L - 107 (*Ancistrus* sp.),
L - 184 (*Ancistrus* sp.) and
L - 172 (*Leporacanthicus* cf. *galaxias*).

Brood caring in loricariids varies, as far as it is known at all, very much.

Some breed in caves, some in bamboo canes so small that they can just get inside, others spawn on roots or plants or other substrates.

Just as variable is the rearing of the offspring: sometimes the eggs are guarded, and some parents neglect their brood completely.

Some fry hatch as quite large young fish, others are so small that they have to be fed with the finest possible food; in some species, the fry grows up pretty fast, in others they grow slowly and for a very long time.

Breeding:
Rearing the young

To give you an example and maybe raise your interest in trying to breed your own catfish offspring, we print this breeding report by Mr. Karl Lang, who kindly provided this article.

Breeding report of

L 107 (or L 184)✳

by Karl Lang †

In May 1996, after more than one and a half years of keeping it, I one day discovered the first spawn of the quite rare *Ancistrus* sp..

Almost 50 eggs were guarded by the male; they had been spawned into a cave made of slate (length 14 cm, width 6 cm, height 4 cm).

The only information I had about the place where the fish had been collected was that they came from the Rio Negro in Brazil.

Usually, fishes coming from this region prefer soft, acidic water.

This piece of information turned out be of great importance for the successful keeping and especially breeding of my loricariids.

Males can be distinguished from females by the clearly visible head bristles, which are only weakly developed in females.

The attractive features of these cats are the bright white, quite large spots on the black body and the caudal fin with its filamentous extensions on the upper and lower end.

Males attain lengths around 12 cm, females are smaller with about 10 cm.

The males stay hidden in the cave during the day, while the females keeps close to the surrounding area of the cave.

Ripe females try continuously to get inside the cave where the male of their choice hides.

The male defends his territory rigorously, so that the females look quite bad after such an attack. When I discovered the eggs, they must have been several days old, because I could already see a light pole in the eggs which had then about 6 mm in diameter.

The eggs were attached in a grape-like lump to the bottom of the breeding cave and were milky white.

Because I knew from a friend that his eggs had been eaten by the male, I decided to hatch them separately.

Still attached to the bottom plate of the cave, I moved the eggs to a simple 2 l tank.

The water was conditioned so that it had the same parameters as the parents' tank (pH 5.5, hardness below measuring, temperature 26°C).

To prevent the eggs from contracting a fungal infection, I installed a gentle air stream and added three alder cones to the water.

After four days I discovered the first fry; it took six days until all eggs had hatched.

The majority of the eggs needed artificial hatching: I stroked them with a fine brush until the shells burst open.

The yolk sacs the larvae of these species have are even larger than the yolk sacs of *Hypancistrus zebra* (L-046).

During the first 3 to 4 days of their lives, the larvae of *Ancistrus* sp. cannot be distinguished from the fry of *Hypancistrus zebra*.

From the fourth day on, you can see that the developing stripe pattern differs.

After 18 days, the large yolk sac had been absorbed in all young fish.

As first food, I gave them pieces of soft wood, overgrown with algae, as well as algae-coated plant leaves and stones from other aquaria.

✳ Both numbers are, like many other L-numbers, one and the same fish. In the following text, the fish is simply called *Ancistrus* sp..

Breeding:
Differences in the sexes

L-107 and L-184 look very much alike. It is probably the same species.

I performed a daily 30% water change with fresh water with about same chemistry as the actual aquarium water.

From the sixth week on at a body length of 2.5 cm, the first young began to display the beautiful coloration of their parents.

Please note that during the first week of their lives, the young do not tolerate a pH above 7.

We are very sad that in February 98, our dear friend and experienced aquarist Karl Lang, who wrote this breeding report, died.

In the future, we will have to cope without our friend's knowledge.

In Peckoltia species (Like this L-002) males and females are quite easily distinguished: left-male, right-female.

Only in few species the sexes are as easily told apart as in L-076- they are simply differently coloured.

In Aphanotorulus (like this L-094) the males develop spines on the caudal base during the spawning season, in Rineloricaria (here: R. lanceolata, no L-number) they show the typical "whiskers".

AQUALOG *Special* The most beautiful L-numbers © Verlag A.C.S. GmbH

Breeding:
Differences in the sex

In Pseudacanthicus (here: L-065), males are clearly more elongate and slender than females. Like in almost all loricariid species, the male's snout is longer than the female's.

Ancistrus species (here: L-180) are quite easily sexed. Although the females do have tentacles on their snouts, too, these are clearly more developed in males.

In Panaque and Peckoltia species, the future females and males can be distinguished from each other at quite an early stage, like in this about 4 cm long, young L-204. Males (left) have a longer and less thick caudal base and a longer head.

In adult Panaque of the nigrolineatus-complex (here: L-027c), the males develop very long interopercular odontodes and hooked spines on the anterior pectoral fin spines.

The AQUALOG - System
Information and Description

AQUALOG Lexicon

The **AQUALOG** team has set itself the goal to catalogue all known ornamental fishes of the world – and this task will, of course, take several years, as there are over 40,000 fish species.

Compiling an **AQUALOG**lexicon, we take a certain group of fishes, label all known species with code-numbers, look for the newest results of fish research about natural distribution, features and maintenance of the fishes and try to get the best photographs, often from the most remote parts of the world.

Our ingenious code-number-system labels every species with its own individual code-number which the fish keeps even if a scientific re-naming occurs.

And not only the species gets a number, also each variety, distinguishing locality, colour, and breeding form.

This system makes every fish absolutely distinct for everybody. With it, international communication is very easy, because a simple number crosses almost all language barriers.

This is an advantage not only for dealers, but for hobbyists, too, and thus for all people involved in the aquarium hobby.

Again and again, new fish species are discovered or new varieties bred. Consequently, the number of fishes assigned to a certain group changes constantly and information from available specialist literature is only reliable within certain time limits. Thus, an identification lexicon that is up-to-date today is outdated after as little as one year.

To give aquarists an identification 'tool' that stays up-to-date for many years, we developed our ingenious patented code-number system.

When going to press, our books contain all fishes that are known to that date. All newly discovered or bred species are regularly published as either supplements or as so-called "stickups" in **AQUALOG**news.

These supplementary peel-back stickers can be attached to the empty pages in the back of the books.

As you can see, we provide the latest information from specialists for hobbyists. Over the years, your **AQUALOG** books will 'grow' to a complete encyclopaedia on ornamental fishes, a beautiful lexicon that is never outdated and easy to use.

AQUALOGnews

AQUALOGnews is the first international newspaper for aquarists, published in four-colour print, available in either German or English language and full of the latest news from the aquatic world.

The following rubrics are included: Top Ten, Brand New, Evergreens, Technics, Terraristics, Fish Doctor and Flora. Further, there are travel accounts, breeding reports, stories about new and well-known fish etc.

The news gives us the opportunity to be up-to-date, because up to one week before going to press, we can include reports and the 'hottest' available information.

This way, every six weeks a newspaper for friends of the aquarium hobby is published that makes sure to inform you about the latest 'arrivals' waiting for you at your local pet shop.

AQUALOGnews can be subscribed to and contains 40 supplementary stickers for your AQUALOG books in 12 issues. You can subscribe to the news either via your local pet shop or directly at the publishers.

Issues without stickups (print run: 80,000) are available at well-sorted pet shops. The newspaper also informs you about newly published supplements.

AQUALOG Special

The *Specials* series is not intended to repeat all the things that were already known twenty years ago, like 'how to build your own aquarium' – something, probably nobody practises anymore, because there is no need to do so.

We provide the latest and most important information on fish keeping and tending: precisely and easily understandable.

We want to offer advice that helps you to avoid mistakes – and help your fishes to live a healthy life.

We intend to win more and more friends for our beautiful and healthy (because stress-reducing!) hobby.

Order our new free catalogue, where all our previous books and the ones in preparation are shown and described.

Ihr Nachschlagewerk
Your reference work

Vervollständigen Sie Ihr Nachschlagewerk durch weitere Bücher der Aqualog-Reihe:

Complete this reference work with further volumes of the Aqualog series:

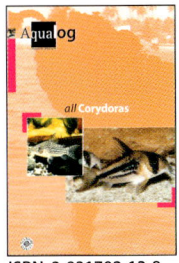

ISBN 3-931702-13-8 
ISBN 3-931702-04-9 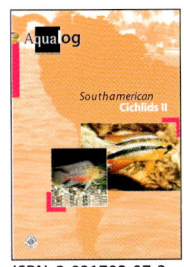
ISBN 3-931702-07-3 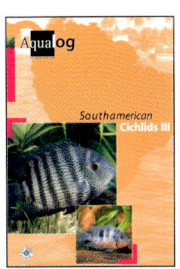
ISBN 3-931702-10-3 
ISBN 3-931702-75-8 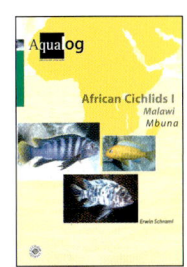
ISBN 3-931702-79-0

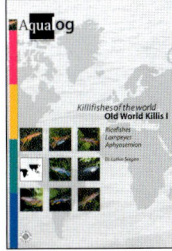

ISBN 3-931702-25-1
ISBN 3-931702-30-8 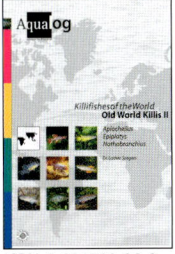
ISBN 3-931702-76-6 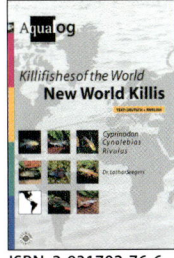
ISBN 3-931702-21-9 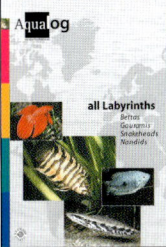
ISBN 3-931702-77-4 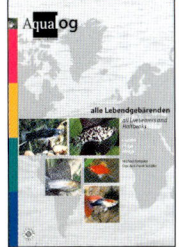
ISBN 3-931702-01-4 

Mehr Informationen direkt bei
For more information please contact

Aqualog Verlag
Liebigstraße 1, D-63110 Rodgau/Germany
Fax: +49 (0) 61 06 - 64 46 92,
email: acs@aqualog.de
Internet: http://www.aqualog.de

ISBN 3-931702-78-2
ISBN 3-931702-93-6 
ISBN 3-931702-80-4 

Alle **Aqualog**-Produkte erhalten Sie im Zoofachhandel und überall auf der ganzen Welt. Wir nennen Ihnen gerne Bezugsquellen.

You can obtain all Aqualog products everywhere in the world. Contact us for addresses.

ISBN 3-931702-61-8
(deutsche Fassung)

ISBN 3-931702-60-X
(English edition)

Aqualog news

Informiert top aktuell zu folgenden Themen:
Up-to-date information on following topics:

- Süß- und Seewasserfische / *marine and freshwater fishes*
- Terraristik / *terraristic*
- Neuentdeckungen / *new discovered fishes*
- Aquarienpflanzen / *plants for the aquarium*

Fordern Sie Ihr Gratis-Exemplar der englischen oder deutschen Ausgabe beim Verlag an!

Order your free specimen copy of the German or English edition at the publisher!

AQUALOG *Special* The most beautiful L-numbers 43

Outlook:
Future L - numbers?

This book already includes pictures of the latest L-numbers, like L-252 and L-256.

But every day, new loricariids are discovered and imported.

They are still labelled with L-numbers, but, unfortunately, following a system nobody understands.

Some new fishes that are offered to the DATZ are only labelled after a very long time, or even not at all.

Also, it happens time and time again, that two different fishes get the same L-number, like, for example, L-241 and L-242.

Instead of correcting this mistake through adding a simple a) or b) to the number in order to distinguish them, the fishes get completely new numbers.

On these three photos you can see three different colorations of L-065

Obviously, the editors of the DATZ do not know, how confusing this situation is for hobbyists and that the DATZ's seemingly illogical labelling policy is hard to understand and accept by aquarists.

To give you an example, how differently these catfishes can look, we show you the L-065, a suckermouth cat that was labelled in the DATZ 7/90 as L-065.

The fish (probably a member of the genus *Pseudacanthicus*) comes from Tocatins (Brazil) and displays such a variable body pattern, that hardly ever two individuals have the same appearance.

For a more detailed report about this fish, please read AQUALOGnews issue No. 13.

Following this chapter, we present some new species to you, species that are waiting (like many others that rest in our archives) to be labelled with an L-number.

Whether they'll get one or not, is as much a mystery to us as it is probably to you.

We could, of course, start to create our own numbers, labelling the many new species we know of with Aqualog-L-numbers...

This would be really easy for us, because we closely work together with our sister-company Aquarium Glaser that has big weekly imports from the natural distribution areas of loricariids.

Further, we co-operate with many fish experts who regularly travel to these regions and always bring new fishes from their trips.

Until today, we have abstained from doing so, because we think it would only add to the already confusing situation - and we intend to do the exact opposite, that is, to help hobbyists to easily find their way through the amazing variety of suckermouth catfishes.

Photos:
New L- numbers on the waiting list

First, one of the many Peckoltia *species. The male is on the left, the female on the right photograph. The species has been imported from Tocatins (Brazil). They look very similar to other* Peckoltia, *but this form is unknown.*

The fish on the left is a Hypostomus *from Paraguay that still does not have an L- or LDA- number. The fish on the right is a* Hypostomus, *too, but it comes from Xingu. This fish is especially interesting, because its coloration is almost identical to that of L -020. The latter comes also from Xingu but belongs to the* Ancistrus *group.*

Another very pretty fish is the new 'zebra' from Xingu, shown on the left-hand photo. If larger imports are successful, this fish will certainly become a 'hit' among hobbyists. The specimen on the right-hand picture, resembling L-041, will definitely remain a subject of admiration by real catfish enthusiasts only.

Also very interesting is the frequently from Peru imported Chaetostoma *species, in which the male has an orange back part whereas the female is ordinarily coloured.*

AQUALOG *Special* The most beautiful L-numbers

Index

L-001	S43001	Glyptoperichthys joselimaianus	"Whitespot Glyptoperichthys"	A-1,10
L-002	S43002	Panaque sp. "Peckoltia vittata" (vermiculata?)	"Tiger Pleco"	B-1,40
L-003	S43003	Panaque (?) sp.		36
L-005	S43005	"Peckoltia angelicus"	"Crow Star"	C-1
L-006	S43006	Peckoltia oligospila	"Brown dot"	D-1
L-007	S43007	Leporacanthicus galaxias	"Rüsselzahnwels /Tooth-Nose"	E-1
L-010a	S43010a	Rineloricaria sp. "red"	"Roter Hexenwels/ Red Lizard Cat"	F-1
L-014	S43014	Scobinancistrus aureatus	"Goldy Pleco"	G-1,29
L-015	S43015	Peckoltia vittata	"Xingu Peckoltia"	H-1
L-017	S43017	Ancistrinae gen. sp.	"Flathead Pleco"	A-2
L-020	S43020	Baryancistrus sp.	"Belem Polka Dot"	B-2
L-021	S43021	Liposarcus cf. pardalis		7
L-025	S43025	Pseudacanthicus cf. serratus	"Scarlet"	21
L-026	S43026	Baryancistrus niveatus	"Niveato Pleco"	C-2
L-027c	S43027c	Panaque sp. nigrolineatus		41
L-030	S43030	Parancistrus sp.	"Peppermint Pleco"	D-2
L-034	S43034	Ancistrus ranunculus	"Kaulquappen Wels" /"Bristle Bushmouth"	E-2
L-046	S43046	Hypostomus zebra	"Zebra Peckoltia"	F-2
L-047	S43047	Baryancistrus sp.	"Magnum Orangeseam Pleco"	G-2
L-050	S43050	Cochliodon sp.	"Tocantins Cochliodon"	H-2
L-051	S43051	Ancistrinae sp.		28
L-060	S43060	Hypancistrus sp.	"Longtail Pleco"	A-3
L-065	S43065	Pseudacanthicus sp. ("variegatus"?)	"Variegated Pleco"	B-3,41,65
L-066	S43066	Ancistrinae gen. sp.	"King-Tiger Pleco"	C-3
L-069	S43069	Ancistrinae gen. sp. (Peckoltia ucayalensis?)	"Pintado Tiger"	D-3
L-070	S43070	Peckoltia sp.	"Zombie Plecko"	E-3
L-071	S43071	Ancistrus sp. "White-spot"	"Tocantins Hoplogenys"	F-3
L-074	S43074	Panaque (?) sp. (Peckoltia sp.?)	"Ringlet Tiger Pleco"	G-3
L-076	S43076	Peckoltia (?) sp. (Parancistrus sp.?)	"Redseam Tiger Pleco"	H-3,40
L-081	S43081	Baryancistrus sp.	"Xingu orangseam Pleco"	A-4
L-083	S43083	Glyptoperichthys cf. gibbiceps		16
L-089	S43089	Ancistrus sp. (tamboensis?)	"Tambo Ancistrus"	B-4
L-090	S43090	Panaque sp. "Papa"	"Papa Panaque"	C-4
L-091	S43091	Leporacanthicus triactis	"Redfin Blackspot"	D-4
L-094	S43094	Aphanotorulus ammophilus	"Golden Cochliodon"	E-4,40
L-096	S43096	Pseudacanthicus sp.	"Blackspot Pleco"	F-4
L-100	S43100	Ancistrus sp.		36
L-102	S43102	Ancistrinae gen. sp. (Peckoltia sp.?)	"Snowball Peckoltia"	G-4
L-103	S43103	Peckoltia (?) sp. (vittata?)	"Hairy Peckoltia I"	H-4
L-106	S43106	Ancistrinae gen. sp. (Peckoltia sp. ?)	"Orangeseam Cat"	A-5
L-107	S43107	Ancistrus sp.		40
L-110	S43110	Ancistrus sp.	"Orange Spotted Pleco"	B-5,14
L-114	S43114	Pseudacanthicus cf. leopardus	"Leopard Acanthicus"	C-5,24
L-121	S43121	Peckoltia cf. platyrhyncha	"Wormline Peckoltia"	D-5
L-124	S43124	Ancistrinae gen. sp. (Hypostomus sp. ?)	"Bigspot Hypostomus"	E-5
L-128	S43128	Ancistrinae gen. sp. (Chaetostoma sp. ?)	"Small Spotted Cat"	F-5
L-137	S43137	Cochliodon cochliodon (?)	"Violet Red Bruno"	G-5
L-139	S43139	Cochliodon sp. (oculeus?)	"Spotted Cochliodon"	H-5
L-157	S43157	Ancistrus (?) sp.	"Red Spotted Ancistrus"	A-6
L-160	S43160	Pseudacanthicus cf. spinosus		28
L-162	S43162	Peckoltia sp.		15
L-163	S43163	Ancistrinae sp.		28
L-168	S43168	Lasiancistrus pictus (?) (Peckoltia?)	"Brazil Butterfly"	B-6
L-174	S43174	Hypancistrus (?) sp. (Hemiancistrus?)	"Black White Ancistrus"	C-6,36
L-180	S43180	Ancistrus sp.		41
L-184	S43184	Ancistrus sp.		40
L-185	S43185	Ancistrinae sp.		33
L-186	S43186	Pseudacanthicus (?) sp.		15
L-188	S43188	Chaetostoma sp.	"White Spotted Chaetostoma"	D-6,16,21
L-190	S43190	Panaque sp. (cf. nigrolineatus)	"Goldenline Royal Pleco"	E-6
L-191	S43191	Panaque sp. (cf. nigrolineatus)	"Brokenline Royal Pleco"	F-6
L-200	S43200	Ancistrinae sp.		28
L-201	S43201	Ancistrinae gen. sp.	"Spotted Black Ancistrus"	G-6
L-203	S43203	Panaque sp. "Ucayali"	"Ucayali Panaquae"	H-6
L-204	S43204	Panaquae sp. "Peru"	"Small Line Peru Panaquae"	A-7,41
L-234	S43234	Megalancistrus cf. parananus	"Parana Giant Ancistrus"	B-7

Index

L-239	S43239	Panaque sp.	"Redfin Hypostomus"	36	
L-241	S43241	Ancistrinae sp.	"Cumina Peckoltia"	33	
L-242	S43242	Hypostomus sp.	"Cumina Lithoxus"	C-7	
L-247	S43247	Peckoltia sp. "Cumina"	"Belem Ancistrus"	D-7	
L-256	S43256	Lithoxus (?) sp.	"Golden Black Ancistrus"	E-7	
LDA-002	S43402	Ancistrus(?) sp. "Belem"	"Small Spot"	F-7	
LDA-003	S43403	Ancistrus sp.	"Golden Marble"	G-7	
LDA-005	S43405	Ancistrus (?) sp. (Peckoltia?)	"Mega Clown Peckoltia"	H-7	
LDA-008	S43408	Ancistrus sp. "Mato-Grosso"	"Gold Orangefin Panaquae"	A-8	
LDA-019	S43419	Peckoltia sp.	"Pitbull Pleco"	B-8	
LDA-022	S43422	Panaque sp.	"Orange Longfin Panaque"	C-8	
LDA-025	S43425	Hypostomus (?) sp.	"Peru II Panaque"	D-8	
LDA-027	S43427	Panaque sp. "Orange-long-Fin"	"Big Whitespot Ancistrus"	E-8	
LDA-028	S43428	Panaque sp.	"Rio Sucker Spot"	F-8	
LDA-033	S43433	Baryancistrus (?) sp.		G-8	
LDA-034	S43434	Neoplecostomus sp. "Rio"		H-8	
		Glyptoperichthys gibbiceps		25	
		Ancistrus sp.		14	
		Otocinclus sp."Negros"		21	
		Loricariide gen. sp.		32	
		Peckoltia sp. "Tocantins"		45	
		Hypostomine sp.		45	
		Ancistrine sp.		45	
		Chaetostoma sp.		45	

Literature

Magazines:

AQUALOGnews
Verlag A.C.S. GmbH
ISSN 1430-9610

DAS AQUARIUM
Birgit Schmettkamp, Verlag
ISSN 0341-2709

DATZ
Die Aquarien- und
Terrarien-Zeitschrift
Verlag Eugen Ulmer
ISSN 0941-8393

DATZ - Sonderheft
Harnischwelse
Verlag Eugen Ulmer

Books:

BASSLEER, G. (1996):
Bildatlas der Fischkrankheiten im
Süßwasseraquarium
Augsburg

FRANKE, H.-J. (1985):
Handbuch der Welskunde
Hannover

GLASER, U. (1995):
Loricariidae all L-numbers
Mörfelden-Walldorf

REICHENBACH-KLINKE, H.-H. (1975):
Krankheiten und Schädigungen
der Fische
Stuttgart

ISBRÜCKER, I. H. J. (1980):
Classification and catalogue
of the mailed Loricariidae
(Pisces, Siluriformes).
Inst. Tax. Zool. Univ. Amsterdam

UNTERGASSER, D. (1989):
Krankheiten der Aquarienfische
Stuttgart

Symbols

In order to include as many pictures as possible, and bearing the international nature of the publication in mind, we have intentionally decided against detailed textual descriptions, replacing them by international symbols. This way, one can easily obtain the most important facts about the species and its care.

Continent of origin:

simply check the letter in front of the code-number
A = Africa   E = Europe + North America
S = South America X = Asia + Australia

Age:

the last number of the code always stands for the age of the fish in the photo:

1 = small   (baby, juvenile colouration)
2 = medium   (young fish / saleable size)
3 = large   (half-grown / good saleable size)
4 = XL   (fully grown / adult)
5 = XXL   (breeder)
6 = show   (show-fish)

Immediate origin:

W = wild
B = bred
Z = breeding-form
X = cross-breed

Size:

..cm = approximate size these fish can reach as adults.

Sex:

♂ male   ♀ female   ♂♀ pair

Temperature:

◁ 18-22°C (68 - 72°F) (room-temperature)
▷ 22-25°C (71 -77°F)   (tropical fish)
△ 24-29°C (75 - 85°F) (Discus etc)
▽ 10-22°C (50 - 72°F) cold

pH-Value:

₽   pH 6,5 - 7,2 no special requirements (neutral)
↓P pH 5,8 - 6,5 prefers soft, slightly acidic water
↑P ph 7,5 - 8,5 prefers hard, alkaline water

Lighting:

○ bright, plenty of light / sun
◐ not too bright
● almost dark

Food:

☺ omnivorous / dry food, no special requirements
☺ food specialist, live food/ frozen food
☹ predator, feed with live fish
☻ plant-eater, supplement with plant food

Swimming:

▣ no special characteristics
⬆ in upper area / surface fish
⬇ in lower area / floor fish

Aquarium- set up:

▭ only floor and stones etc.
▭ stones / roots / crevices
▭ plant aquarium + stones / roots

Behaviour / reproduction:

♥ keep a pair or a trio
≋ school fish, do not keep less than 10
⊱ egg-layer
⊱ livebearers / viviparous
⊶ mouthbrooder
⊂⊃ cavebrooder
⚶ bubblenest-builder
○ algae-eater / glass-cleaner (roots + spinach)
◇ non aggressive fish, easy to keep (mixed aquarium)
⚠ difficult to keep, read specialist literature beforehand
🛑 warning, extremely difficult, for experienced specialists only
❶ the eggs need a special care
§ protected species (WA), special license required ("CITES")

Minimum tank: capacity:

ss	super small	20 - 40 cm	5 - 20 l
s	small	40 - 80 cm	40 - 80 l
m	medium	60 - 100 cm	80 - 200 l
L	large	100 - 200 cm	200 - 400 l
XL	XL	200 - 400 cm	400 - 3000 l
XXL	XXL	over 400 cm	over 3000 l

(show aquarium)

Inches
0 1 2 3 4 5 6 7 8
Centimeter

A

S43001-5　L 1 *Glyptoperichthys joselimaianus* Weber, 1991
"WHITESPOT-GLYPTOPERICHTHYS"
Terra typica: Aruana, Goiás　　　　　　　　　DATZ 12/88
Tocantins (Brazil), „direction Tucuruí", W, 25-30cm
Photo: E.Schraml

S43017-4　L 17 Ancistrinae gen. sp.
"FLATHEAD - PLECO"　(similar to L 17 + L 67)
DATZ 3/89
Xingu near Altamira/Brazil, W, 12-15cm
Photo: F.Teigler/A.C.S.

S43060-3　L 60 *Hypostomus* sp.
"LONGTAIL - PLECO"
DATZ 3/90
Oiapoque-System (Brazil/French-Guyana), W, 20-25cm
Photo: E. Schraml/A.C.S.

S43081-4　L 81 *Baryancistrus* sp.
"XINGU ORANGESEAM - PLECO" (very similar to

B

S43002-4　L 2 *Panaque* sp. "Peckoltia vittata" (vermicula
"TIGER-PLECO" (similar to L 8)
DATZ 1
Tocantins near Cametá (Brazil), W, 8-12cm

S43020-1　L 20 *Baryancistrus* sp.
"BELEM POLKA-DOT"
Adult: kleinere Tüpfel/smaller spots　　DATZ 3/89 + 1
Xingu near Altamira/Brazil, W, 18-22cm

S43065-2　L 65 *Pseudacanthicus* sp. ("variegatus"?)
"VARIEGATED - PLECO"
DATZ 7
Tocantins/Brazil (?), W, 20-30cm
F.Teigle

S43089　L 89 *Ancistrus* sp. (*tamboensis*?)
"TAMBO - ANCISTRUS" (ähnlich/similar to L 156)

C

S43005-4 L 5 "Peckoltia angelicus"
"GROW - STAR" (sehr ähnlich/very similar to L 4)
DATZ 12/88
Tocantins near Cametá (Brazil), W, 8-10cm
Photo: E.Schraml

S43026-3 L 26 *Baryancistrus niveatus* (CASTELNAU, 1855)
"NIVEATO - PLEO"
DATZ 5/89
Tocantins + Araguaia/Brazi, W, 30-40cm
Photo: U.Werner

S43066-3 L 66 Ancistrinae gen. sp.
"KING - TIGER - PLECO"
DATZ 9/90
lower Xingu- and Tocantins-Area, W, 8-12cm
Photo: F.Teigler/A.C.S.

S43090-4 L 90 *Panaque* sp. "Papa"
"PAPA - PANAQUE"
DATZ 1/92

D

S43006 L 6 *Peckoltia oligospila* (GÜNTHER, 1864)(?)
"BROWN - DOT"
DATZ 12/88
Rio Guamá near Ourém (Brazil), W, 8-10cm
Photo: F.Teigler / ACS

S43030-3 L 30 *Parancistrus* sp.
"PEPPERMINT - PLECO" (sehr ähnlich/very similar to L 31 + LDA 4)
DATZ 7/89
Tocantins near Marabá/Brazil, W, 8-10cm
Photo: E.Schraml

S43069-3 L 69 Ancistrinae gen. sp. (*Peckoltia ucayalensis*?)
"PINTADO - TIGER" (similar to L 8)
DATZ 10/90
Tapajós-Area near Santarem/Brazil, W, 10-12cm
Photo: H.G.Evers

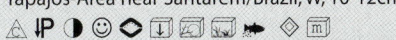

S43091-3 L 91 *Leporacanthicus triactis* ISBRÜCKER, NIJSSEN & NICO, 1993
"REDFIN BLACKSPOT"
terra typica: Rio Mavaca
Orinoco/Venezuela + Puerto Inirida/Columbia, W, 18-22cm
DATZ 4/92

E

S43007-4 L 7 *Leporacanthicus galaxias* ISBRÜCKER & NIJSSEN, 1989
"RÜSSELZAHNWELS / TOOTH-NOSE" (similar to L 172 + L 29)
Juvenil: oft mit gelben Punkten / yellow spots possible   DATZ 12/88
Rio Guamá+Jamari/Brazil, W, 20-25cm
Photo: U. Werner

S43034-4 L 34 *Ancistrus ranunculus*
MULLER, RAPP PY-DANIEL & ZUANO, 1994
"KAULQUAPPEN-WELS / BRISTLE-BUSHMOUTH"   DATZ 7/89
Xingu+Tocantins/Brazil, W, 10-15cm
Photo: E. Schraml

S43070-3 L 70 *Peckoltia* sp.
"ZOMBIE - PLECO"
DATZ 10/90
Rio Tapajós-System, near Santarém/Brazil, W, 10-15cm
Photo: E. Schraml

S43094-2 L 94 *Aphanotorulus ammophilus* ARMBRUSTER & PAGE, 1996
"GOLDEN - COCHLIODON"
DATZ 6/92

F

S43010a-3 L 10a *Rineloricaria* sp. "red"
"ROTER HEXENWELS / RED LIZARD-CAT"

nur Nachzucht/wild unknown, B, Z, 10-15cm
Photo: E. Schraml

S43046-4 L 46 *Hypancistrus zebra* ISBRÜCKER & NIJSSEN, 1991
"ZEBRA - PECKOLTIA"
Bereits nachgezogen/successfully bred   DATZ 9/89 + 6/95
Xingu, oberhalb/upstream of Altamira/Brazil, W, 8-12cm
Photo: E. Schraml

S43071-4 L 71 *Ancistrus* sp. "white spot"
"TOCANTINS - HOPLOGENYS"
DATZ 10/90 + 9/94
Tapajós-Area, near Santarem/Brazil, W, 10-15cm
Photo: E. Schraml

S43096 L 96 *Pseudacanthicus* sp.
"BLACKSPOT - PLECO"
DATZ 6/92

G

S43014-1 L 14 *Scobiancistrus aureatus* Burgess, 1994
"GOLDY - PLECO" (cf. *pariolispos?*)   JUVENIL
DATZ 4/95 + 8/95
Xingu near Altamira/Brazil, W, 30-45cm
Photo: E.Schraml

S43047-3 L 47 *Baryancistrus* sp.
"MAGNUM ORANGESEAM - PLECO"
DATZ 10/89
Rio Xingu/Brazil, W, 20-25cm
Photo: E.Schraml

S43074 L 74 *Panaque* (?) sp. (*Peckoltia* sp. ?)
"RINGLET - TIGER - PLECO"
DATZ 11/90
Rio do Pará, near Portel, Delta of Amazonas/Brazil, W, 6-8cm
Photo: F.Teigler/A.C.S.

S43102-4 L 102 Ancistrinae gen. sp. (*Peckoltia* sp.?)
"SNOWBALL - PECKOLTIA"
DATZ 7/92

H

S43015-3 L 15 *Peckoltia vittata* (Steindachner, 1882)
"XINGU - PECKOLTIA"
DATZ 3/89
Xingu near Altamira/Brazil, W, 7-10cm
Photo: E. Schraml/A.C.S.

S43050-2 L 50 *Cochliodon* sp.
"TOCANTINS - COCHLIODON"   (ähnlich/similar to L 167)
DATZ 10/89
Xingu or Tocantins/Brazil, W, 12-15cm
Photo: Archiv A.C.S.

S43076-3 L 76 *Peckoltia* (?) sp. (*Parancistrus* sp. ?)
"REDSEAM TIGER - PLECO"
DATZ 11/90
Rio do Pará, near Portel, Delta of Amazonas/Brazil, W, 10-15cm
Photo: F.Teigler/A.C.S.

S43103-4 L 103 *Peckoltia* (?) sp. (*vittata?*)
"HAIRY - PECKOLTIA I"
DATZ 8/92

Poster FP 58

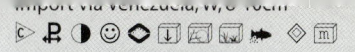

© Verlag A.C.S. GmbH Rothwiesenring 5 64546 Mörfelden-Walldorf Fax: +49 (0)6105 - 64 46 92

43433-3 LDA 33 *Baryancistrus* (?) sp.
"BIG WHITESPOT - ANCISTRUS"
"Das Aquarium" 7/97
ao Luis, Rio Tapajós, Estata Para/Brazil, W, 15-20cm
Photo: F.Warzel

S43434-3 LDA 34 *Neoplecostomus* sp. "Rio"
"RIO SUCKER-CAT"
"Das Aquarium" 10/97
Braucht kühles Wasser/needs cool water
S-Brazil, W,10-12cm
Photo: E.Schram/A.C.S.

43403-4 LDA 3 *Ancistrus* sp.
"GOLDEN-BLACK ANCISTRUS"
"Das Aquarium" 11/92
Brazil, W, 8-10cm
Photo: E.Schraml

S43405-3 LDA 5 *Ancistrus* (?) sp. (*Peckoltia* ?)
"SMALL - SPOT"
(ähnlich/similar to L 136a)
"Das Aquarium" 11/92
Brazil, W, 8-10cm
Photo: F.Teigler/A.C.S.

43201-4 L 201 Ancistrinae gen. sp.
"SPOTTED BLACK-ANCISTRUS"
DATZ 9/95
Venezuelian border with Brazil, W, 8-12cm
Photo: E.Schraml/A.C.S.

S43203-3 L 203 *Panaque* sp. "UCAYALI"
"UCAYALI - PANAQUE"
DATZ 2/96
Rio Ucayali/Peru, W, 20-25cm
Photo: E.Schram/A.C.S.

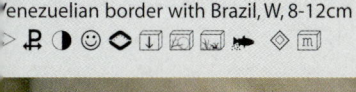

43137-3 L 137 *Cochliodon cochliodon* (?)
"VIOLET RED BRUNO" (sehr ähnlich/very similar to L 50)
DATZ 9/93
apajós-Area/Brazil, Paraguay, W, 15-20cm
Photo: U.Werner

S43139-3 L 139 *Cochliodon* sp. (*oculeus* ?)
"SPOTTED COCHLIODON"
DATZ 9/93
Tapajós-Area/Brazil, Rio Negro/Brazil(?), W, 15-20cm
Photo: F.Teigler/A.C.S.

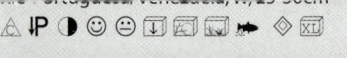

S43124-2 L124 Ancistrinae gen. sp. (Hypostomus sp.?)
"BIGSPOT – HYPOSTOMUS" (ähnlich/similar to L 75 + LDA 2)
DATZ 5/93
Import via Venezuela, W, 20-25cm
Photo: E.O. v.Drachenfels

S43190-3 L190 Panaque sp. (cf. nigrolineatus)
"GOLDENLINE – ROYAL-PLECO"
DATZ 2/95
Columbia, W, 25-33cm
Photo: E.Schraml

S43256-3 L256 Lithoxus (?) sp.
"CUMINA – LITHOXUS"
DATZ 1/98
Rio Cuminá, Pará/Brazil, W, 8-12cm
Photo: Ch. Seidel

S43427-3 LDA 27 Panaque sp."Orange long Fin"
"ORANGE LONGFIN – PANAQUE"
"Das Aquarium" 9/96
Peru, W, 13-16cm
Photo: E.Schraml/A.C.S.

S43128-3 L128 Ancistrinae gen. sp. (Chaestostoma sp.?)
"SMALL SPOTTED CAT" (ähnlich/similar to L 200)
DATZ 5/93
Import via Venezuela, W, 10-12cm
Photo: F.Teigler/A.C.S.

S43191-3 L191 Panaque sp. (cf. nigrolineatus)
"BROKENLINE – ROYAL-PLECO" (ähnlich/similar to L 27)
DATZ 2/95
Columbia, W, 25-33cm
Photo: E.Schraml

S43402-3 LDA 2 Ancistrus (?) sp."BELEM"
"BELEM – ANCISTRUS" (ähnlich/similar to L 75 + L 124)
"Das Aquarium" 11/92
Brazil, W, 10-12cm
Photo: E.Schraml

S43090b-3 LDA 28 Panaque sp.
"PERU II – PANAQUE" (= L 90b)
farbwechselnd/colour changing
"Das Aquarium" 3/97
Peru, W, 25-30cm
Photo: E.Schraml

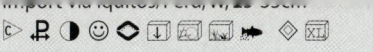

S43114-3 L 114 *Pseudacanthicus* cf. *leopardus*
"LEOPARD - ACANTHICUS" (sehr ähnlich/very similar to LDA 7)
DATZ 10/92
Rio Negro/Brazil, W, 25-30cm
Photo: F.Teigler/A.C.S.

S43174-3 L 174 *Hypancistrus*(?) sp. (*Hemiancistrus*?)
"BLACK-WHITE ANCISTRUS"
DATZ 8/94
Rio Xingu near Altamira/Brazil, W, 12-15cm
Photo: F.Teigler/A.C.S.

S43242-3 L 242 *Hypostomus* sp.
"REDFIN - HYPOSTOMUS"
DATZ 11/97
Rio Orinoco near Rio Atabapo, W, 15-22cm
Photo: F.Teigler/A.C.S.

S43422-3 LDA 22 *Panaque* sp.
"GOLD ORANGEFIN PANAQUE"
"Das Aquarium" 9/95
Venezuela, W, 8-10cm
Photo: E.Schraml/A.C.S.

S43121-4 L 121 *Peckoltia* cf. *platyrhyncha* (FOWLER, 1943)
"WORMLINE - PECKOLTIA"
DATZ 4/93
Import via Guyana, W, 10-13cm
Photo: E.Schraml

S43188-3 L 188 *Chaetostoma* sp.
"WHITE SPOTTED CHAETOSTOMA" (ähnlich/similar to L 187)
DATZ 10/94
Orinoco near Puerto Ayacucho, Venezuela, Columbia, W, 12-15cm
Photo: E.Schraml

S43247-3 L 247 *Peckoltia* sp. "CUMINÁ"
"CUMINÁ - PECKOLTIA" (sehr ähnlich/very similar to L 218)
Juvenil: andere Zeichnung/different coloration DATZ 12/97
Rio Cuminá/Brazil, W, 8-12cm
Photo: Ch. Seidel

S43425-3 LDA 25 *Hypostomus* (?) sp.
"PITBULL - PLECO"
"Das Aquarium" 11/95
Brazil, W, 6-8cm
Photo: F.Teigler/A.C.S.

5

S43106-4 L 106 Ancistrinae gen. sp. (Peckoltia sp.?)
"ORANGESEAM - CAT" (ähnlich/similar to L 122)
DATZ 8/92
Import via Venezuela, W, 10-12cm
Photo: E.Schraml/ACS

S43110-4 L 110 Ancistrus sp.
"ORANGE SPOTTED PLECO" (ähnlich/similar to L 157)
DATZ 9/91
Brazil, W, 8-10cm
F.Teigler

6

S43157-3 L 157 Ancistrus (?) sp.
"RED SPOTTED ANCISTRUS" (ähnlich/very similar to L 110)
DATZ 3/94
Rio Negro near Barcelos/Brazil, W, 8-10cm
Photo: E.Schraml

S43168-3 L 168 Lasiancistrus pictus (?) (Peckoltia?)
"BRAZIL - BUTTERFLY" (sehr ähnlich/very similar to L 5
DATZ 7/9
Rio Negro/Brazil, W, 10-12cm
U.

7

S43204-2 L 204 Panaque sp."PERU" semi-adult
"SMALL LINE PERU-PANAQUE"
DATZ 2/96
Rio Ucayali/Peru, W, 15-20 cm
Photo: Ingo Seidel

S43324-3 L 234 Megalancistrus cf. parananus
"PARANA - GIANT-ANCISTRUS"
DATZ 11/
Rio Paraná/Brazil, W, 40-60cm
E.Schraml

8

S43408-4 LDA 8 Ancistrus sp."MATO-GROSSO"
"GOLDEN - MARBLE" (ähnlich/similar to L 104)
"Das Aquarium" 3/93
Brazil, W, 7-8cm
Photo: F.Teigler/A.C.S.

S43419-4 LDA 19 Peckoltia sp.
"MEGA-CLOWN PECKOLTIA" (ähnlich/similar to L 12
"Das Aquarium" 1/9
Venezuela, W, 8-10cm
E.S